Sub-quantum:

A no brane-r.

Jan Venters

Contents

"Everything is energy, and that is all there is to it"

Albert Einstein

INTRODUCTION

In order to understand everything about our universe, we also have to understand everything about what is not our universe. We must grasp infinity before a theory of everything can arise, for everything includes infinity.

In a finite world, all language is based on a finite view. Big, small, area, volume, period; these are all words with a meaning in a finite world, but cannot be applied to physical infinity. This causes distinct problems in talking of the infinite, as we have no words to describe it and no maths to measure it.

Our maths was developed in a 4 dimensional, finite world so cannot cope with infinity or the one or two dimensional entities as proposed in this hypothesis. The scientific method cannot be used as a method of proof of the infinite. Infinity cannot be measured, so numbers, and consequently mathematics as a whole, cannot be applied, despite its appearance (but possibly not acceptance) in many branches of mathematics.

A Grand Unified Theory, or theory of everything, must include everything. Infinity is everywhere in science. It must be included in a GUT, despite mathematicians' reluctance.

As this hypothesis is based on logic and philosophy, not mathematics, a Grand Unified Theory can be hypothesised, arising directly from the basic logical premise of this thesis, (that the universe is created from a single fundamental entity and a single force arising out of infinity) but cannot be proven by the Scientific Method. Consequently, this hypothesis can never become a theory in the accepted sense based on the Scientific Method as the best proof available is 'true beyond all reasonable doubt'. However, this basis for proof is good enough for a court of law. Is it good enough for the court of my peers?

Can this hypothesis, elegant and easily understood as it is, be accepted as the basis for further research, or will it be rejected out of hand as too simplistic, flying in the face of much currently accepted, mathematically 'proven' but paradoxical, theory as it does? Some such mathematically 'proven' theories accept massless particles as true, while conveniently forgetting that $E=mc^2$ which can be physically measured as true. A massless particle must therefore also have zero energy but are imagined as moving. Can they really be accepted as 'proven'? Current light theory is also not satisfactory in several ways. It seemed to me that a 4 dimensional particle, the photon, does not acceptably explain how light behaves including wave/particle duality, not least because we do not see that in any larger structures. Waves decay too rapidly to be part of such a fundamental as light. 4 dimensional particles flowing in beams must surely block beams flowing at right angles, so how do we see across them? Similarly waves would interfere with each other across the vast reaches of space; stars would be unobservable. We cannot even see the moon reflected on a choppy sea so how could we see pinpoint stars many light years away if their light is travelling in waves! Also, as above, zero mass and zero energy particles, such as photons under the standard model, cannot exist otherwise the speed of light would be zero too when $E=mc^2$ is considered.

I therefore went back to basics and thought about what a fundamental entity (it must be too small to be a particle) should be like. This did not agree with the standard model, and led to a whole new area of thought concerning how light and matter could be created from pure energy, which is what I believe we started with at the Big Bang. The standard model's 17 different particles and 4 different forces all coming into existence at once just seems too much, especially as the start is envisaged as arising from a dimensionless singularity, but these particles are 4 dimensional. This hypothesis describes how the universe is logically possible if created from just one point entity and one force. One then has to consider where that entity came from, of course. The only possible answer is infinity. We know that energy cannot be created or destroyed, so it must be infinite and exist beyond this finite universe. Which meant I had to define what I was talking about, as infinity has many versions, none of which is well understood.

In particular, mathematical logic allows infinity to have a set. But, it seems to me, true infinity can have no boundaries while all sets have at least one boundary; a start boundary. Usually that boundary is zero, but it is still a boundary. Although such a set is illogical, and mathematical infinity has had to be adapted to accept it, most physicists believe it to be true. Hence the many paradoxes arising at quantum level.

The basic hypothesis of a physically infinite fundamental entity laid out in the third of the 7 propositions following can be expanded by simple, philosophically logical processes to provide not only a new theory of light which explains the twin slit experiment without the need for wave/particle duality, but also a Grand Unified Theory, (or Hypothesis), elegantly explaining black holes, dark energy, dark matter, action at a distance and the origin of the universe, et al._

Crucially, patterns and arrangements seen in larger, cosmological structures neatly repeat at sub-atomic level, due to gravity, the main property of the mote (my name for the fundamental entity, explained in proposition 3), prevailing throughout without the need for other forces, so overcoming what, until now, has been a major stumbling block to a GUT.

It requires only 1 entity and one force, thereby eliminating the problems associated with the standard model, while allowing existing properties of matter at atomic scale and above to co-exist. The hypothesised new structure of the atom reflects very well structures seen at larger scales throughout the universe, while agreeing in broad outline with the physically observed structure of the atom (as opposed to the mathematically calculated structure), and embraces Heisenberg's uncertainty principle and Pauli's exclusion principle in its description.

This same hypothesis also predicts that, because the universe is on a hyperbolic expansion curve, it will return to infinity, and everything will end much sooner than current predictions of its end.

All this arises simply and elegantly from the acceptance of the possibility of physical infinity, where 4 dimensional maths cannot apply, and the recognition of what that implies.

Our maths is designed to solve the problems of a 4 dimensional universe, and this hypothesis rests on a 1 dimensional entity. Its time dimension cannot be infinite as it has a start boundary; the Big Bang. However, its spacial dimensions can be, and are, physically infinite. That does not mean very large, or very small, but does mean unmeasurable, thereby precluding the application of mathematics.

It seems unlikely that this hypothesis can be proved by mathematics as we know it today. This, of itself, does not necessarily make this hypothesis wrong.

The current view of the universe is something like this, in descending order of magnitude:

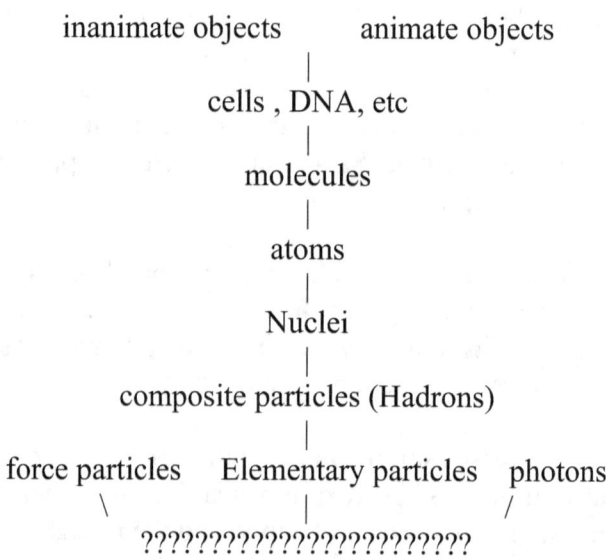

inanimate objects   animate objects
|
cells , DNA, etc
|
molecules
|
atoms
|
Nuclei
|
composite particles (Hadrons)
|
force particles   Elementary particles   photons
\ | /
???????????????????????????

Below the level of elementary particles no-one seems sure of what happens, but they are 4 dimensional and therefore theoretically divisible. And photons suddenly seem to appear from nowhere. The forces acting on these particles are suggested as further particles, (gluons) which makes no sense, because if they are also particles, what force forces the force particles to act on the other particles, and if gluons are particles, how do they act at a distance, as gravity does? Or is there also an undiscovered gravity particle?

There are various theories (String theory; M theory; branes, etc) but none does a good job of unifying all the various theories from other branches of physics. In particular, gravity is a problem, and unifying Newtonian (large scale) and Quantum (small scale) physics is the Holy Grail. The Photon (i.e. light) is lumped in with elementary particles, with no explanation of why we perceive it to be different from them. And how can it be a component of matter when it travels at such speeds? Heat (energy) does not seem to feature at all in the standard model, despite it being present throughout the universe.

This hypothesis accepts the Big Bang theory and all known physics above particle level, but considers the multiplicity of particles and forces indicated by the standard model seem unlikely to have suddenly come into joint existence at the start of everything. It is just too complicated to be the origin of everything. (In fact I still think it too complicated to apply to matter now. See proposition 5 concerning a new proposal for the structure of atoms, despite the apparent 'proof' provided by mathematical manipulation, with constant updating of the formulae, to arrive at the current model. And there are still discrepancies!.) This hypo-thesis proposes a possible explanation of the origins of the universe at sub-particle level with just one en-tity and one force pre-existant. This has the major ad-vantage that the forms that logically arise at sub-atomic level mirror the form of the larger structures found throughout the universe, from atoms to galax-ies.

PROPOSITION 1
THAT PHYSICAL INFINITY
IS A STATE OF ENERGY

In order to construct a hypothesis of everything we have first to consider the definition of infinity. This may seem a little way from a Grand Unified Theory, and has not previously been included in such a thesis but is a necessary first step for a full understanding of the proposal, and to ensure the reader grasps the major differences in understanding that can arise when referring to infinity.

Infinity has many guises: mathematical infinity; religious infinity; logical infinity; cosmological infinity; computing infinity to name but a few. Since the ancient Greeks, mathematicians have tried and failed to prove that infinity actually exists in a mathematical universe. They have had to come up with several flavours of infinity to cover all the different possibilities, depending whether you are talking about calculus, set theory; countable or uncountable numbers; etc; etc.

In order to distinguish it from these others, I shall refer to the physical infinity I am talking about as

(p)infinity. Logically, mathematical infinity then becomes **(m)infinity.**

So what must a physical infinity ((p)infinity) look like? Our language only describes our 4 dimensional universe, so we must deal in negatives and use oxymorons:

(p)infinity can have no boundaries:

It must go on forever, although 'going on' and 'forever' do not apply at infinity as they imply movement, dimensions and time . (p)Infinity simply is.

(p)infinity can have no units:

Units are a mathematical construct to enable measurement. You can always manipulate units (e.g. add 1; subtract 1) which means they always have a start boundary (often 0) and always reach an end boundary, so they cannot be (p)infinite; only have an end boundary of (m)infinitely many. This is only one dimension. Any dimension always has 2 boundaries; a start and an end, even if that end is at (m)infinity. Any object which is mathematically infinitely long has 2 other spacial dimensions, which may or may not be (m)infinite. If these are also (m)infinitely long or short they all converge at a start point (0). Thus a point entity, as later hypothesised, is mathematically possible. If it's time dimension is also zero it becomes a singularity, which is what is at the centre of a black hole, as we shall see. Mathematics can envisage a unit called infinity, and can have multiple infinities. However, if we accept that infinity can have no boundaries, more than 1 (p)infinity is not possible. Consequently, more than 1 entity in that infinity is not possible. (P)infinity just is.

(p)infinity can have no dimensions:

If it has no units, it can have no dimensions, as units are required to measure those dimensions. 37 is meaningless as a dimension without a unit attached. If it has no dimensions it can take up no space.

(p)infinity can have no time:

If it has no dimensions, it can have no time, as time is also a dimension, requiring units (hours, years etc) to measure it in. Any unit of time, however small, would give (p)infinity too long to happen and could be divided. Measurement of Time always has a beginning and an end so cannot be (p)infinite. The beginning of time in this universe was the Big Bang. The end of time has not yet occurred, but it will (see proposition 6).

(p)infinity can have no spacetime:

Obviously, from the above. Consequently it cannot physically exist in this universe, which is fully measurable in spacetime. But it cannot be outside the universe either, as that implies boundaries. It is literally everywhere and nowhere at the same time. It just 'is'.

(p)infinity cannot sustain movement:

It has no units of anything to move, and has no dimensions to move in.

No more than one thing can be in the same (p)infinity:

Any more than one and you have units of each. And that thing just 'is' in that infinity.

However, I shall show that there is one positive attribute to (p)infinity:

INFINITY IS NOT A PLACE, A MEASUREMENT, NOR A NUMBER. IT IS A STATE OF ENERGY.

Dimensions have to be measured in units of some kind. This includes time, the 4th dimension. From this one can argue, as above, that (p)infinity, which has no dimensions or units, is not a number, nor a space, nor a time; not a measurement at all but a state. A (p)infinite entity is not 'in infinity', it is just 'infinite', just as it could be 'solid' or 'liquid'.

It can change phase just as the state of water can change from liquid to solid. It is only after this phase

17

transition that entities can then take on boundaries, dimensions and other physical properties. Mathematics can be applied. And, most importantly, as we shall see, it has room to move, allowing time to be measured.

I again stress that I deal here with a view of physical infinity from a philosophical standpoint, while considering problems in cosmological, quantum and other real life fields. This philosophical approach is necessary, as maths cannot be applied wherever (p)infinity also applies.

I believe that a physical infinity ((p)infinity) exists, but not in a 4 dimensional mathematical universe. It can be demonstrated by logical argument, but is outside the realm of mathematics, pure or applied. Applied mathematics needs an approximation of infinity, something that is 'infinity-like' to solve some of its problems, but this is not infinity pure and simple and causes many paradoxes. It is, in effect, (m)indefinity rather than (m)infinity. Its numerical set is defined in Wikipedia to be $\{0...(\infty-1)\}$. Having a set means it has boundaries, so cannot be a true infinity.

As an aside, it seems to me that if numerical infinity were to have a set it would be $\{(\infty+0)...(\infty-0)\}$, (The upper limit cannot be $(\infty-1)$; what about $(\infty-0.1)$ or $(\infty-0.00001)$ or any smaller deduction?) but then I am not a mathematician. I shall show later that (p)infinity is equivalent to but not equal to zero.

An entity, set or measurement cannot expand or contract into (p)infinity; one can always add 1 or -1 (or any other quantity) to it. It can only be said to change state or phase to (p)infinite. Phases change when a certain amount of energy, usually in the form of heat, is added to or removed from any material, (e.g. heat water enough and it turns to steam; cool it enough and it turns to ice) and this hypothesis argues that everything is made of energy at a basic level. Add or remove enough energy and a phase change will happen to (p)infinity.

I argue that the traditional four fold phase transition sequence (plasma; gas; liquid; solid) should be extended to:

(p)infinite: plasma: gas: liquid: solid: plasma (?) (Bose-Einstein condensate?): (p)infinite

This phase sequence applies to all elements and simple molecules, and is solely due to the addition or removal of energy. Any more complicated structure will break down into its constituent parts before they in turn take a phase transition.

I will argue later that everything in this universe is made of units of energy, and they must always be moving, to enable time to be measured. That their basic speed happens to be the speed of light is a consequence of the mass of the basic unit, which I call the mote (See proposition 3 in this series).

We know that the addition of more and more energy, usually in the form of heat, will turn a solid into a liquid and then into a gas. Changes in pressure or other forms of energy will change the heat energy needed, although the overall energy will be the same. More energy will break down the molecules into atoms and then into their constituent parts to form a plasma. I contend that even more energy would turn that plasma into (p)infinity, pure energy, which is not observable or measurable. It does not move and has no time. The explanation for this will occur in proposition 3 of this series where I propose a physical nature for energy.

At the other end of the scale, removal of energy results in the phase transition from a liquid to a solid and then, as absolute zero (i.e. the complete absence of energy) is approached, first the molecules and then the atoms lose their energy, allowing superfluidity and superconductivity Remove more energy and eventually, molecules and atoms, then elementary particles, break into their constituent parts so that their energy can be removed; as absolute zero is reached, there would be a complete absence of energy in the physical space, a phase transition will occur and because everything is made of energy (as I shall show in proposition 3 of this series), which cannot be destroyed, nothing is left except a pre-existing (p)infinity of energy, which cannot be observed.

Solids reduced by the removal of energy to absolute zero will disappear as absolute zero means zero energy and no substance with zero energy can exist, as particles, even within atoms, are always moving, as is everything in the universe.

However, even if all matter in the universe were to disappear, as I will show it eventually will, there would still be (p)infinite energy, as energy cannot be destroyed. (P)infinity is therefore equivalent, but not equal, to nothing or zero.

Temperatures (energy level) at which phase transitions happen vary between materials, and at different pressures, but all will follow this route if enough energy is applied or removed. In most cases this will occur outside the range of current technology.

It might be surmised that the removal of further energy from all solid materials will result in a similar phase state to plasma, as results from the addition of energy at the other end of the scale, but technology is not sufficiently advanced to achieve removal of energy to such low levels, approaching very close to absolute zero, so this cannot be demonstrated. However, logic suggests that such is the case, as, after all free energy has been removed, the atoms must be broken down to remove the energy from which they are made. In effect, if absolute zero were reached, a black hole would be created (See proposition 4 of this series).

Prediction: If absolute zero were ever approached by some future technology, the temperature would suddenly rise just before it was attained as the huge number of motes (see proposition 3 of this series) in the nuclei of the atoms was released in a similar way to what occurs in a nuclear explosion, although energy levels at that stage would be so low that such extreme violence would not happen.

At this point, when all heat/kinetic energy has been removed, the total energy of the material is reduced to zero. However, energy cannot be created or destroyed, so this must also mean that while the material would disappear at absolute zero, the energy must remain in some form. If Einstein is to be believed, if $E=mc^2$, at zero energy levels, mass and speed must also equal zero. In all atoms, particles are supposed as moving around and within the nucleus. If there is no energy present, they cannot move, so the atoms would disintegrate. An entity with zero mass and zero energy cannot be observed, and therefore cannot be measured. Its speed will also be zero. This means it is either physically nothing (has miraculously disappeared) or has changed phase state to (p)infinite.

Bose-Einstein condensate is a low temperature effect, acting on a very low density gas, where the phase transitions to liquid and solid are not possible due to insufficient density of molecules to create these structures. so can be considered a special case. It may also be the plasma- like phase conjectured above, so I have included it in the phase transition list. Removal of sufficient energy from the Bose-Einstein condensate would result in an immediate jump to the (p)infinite phase.

Those physicists who claim zero mass particles exist must be wrong, as the particles would have zero energy also, so could not move or do anything useful. (or else they are only referring to a mathematical construct to make their equations work, despite Einstein!)

I have heard it suggested that it is possible because $E=mc^2$ is not the full equation. That is $E=mc^2+pv^2$, so if m=0 the equation becomes $E=pv^2$. Unfortunately, p=mv, so this equation would still contain a multiplication by zero.

The photon is considered to have zero mass, but I shall show in proposition 3 of this series that energy acquired mass at the Big Bang by accelerating, and that light is made from pure energy, so must also have mass. It is known that as a star burns it loses mass, resulting in the various forms stars take due to changes in their gravity. They are constantly expelling vast quantities of energy in the form of heat and light, and thereby reducing their mass. They are not losing anything else so heat and light must have mass.

Incidently, if the photon has zero mass, as supposed by the Standard Model, how did it accelerate from the Big Bang during inflation without acquiring it?

(I should mention, before anyone else does, that the phase transition of water to ice is another exception, where ice is less dense than water. This is attributed to the presence of hydrogen bonds which causes atoms to become more distant in the solid phase (*Wikipedia Jan 2018)*. The overall amount of energy is still reduced from liquid to solid, demonstrated by the fact that it gets colder)

Within this universe everything must have at least one boundary. The major of these is the start boundary of the Big Bang, resulting in everything having measurable time. This means that everything that exists in this universe has at least one dimension, time, which started at the Big Bang, so nothing existing in this universe can be wholly (p)infinite. However, this definition of physical infinity is a philosophical one. Nothing I suggest in any way affects existing mathematically proven theories involving mathematical infinity, such as set theory, uncountability and other mathematical theories, as little that follows can be mathematically manipulated or proven, as will become clear. (And as noted above, mathematical infinities are always approximations.) This problem with mathematics, of itself. does not make this hypothesis wrong.

In particular, I will demonstrate the paradox that physical infinity is equivalent to zero, but not equal to it.

Let us assume for the moment that the symbol ∞ represents physical, rather than mathematical infinity. Because physical infinity is not within the sphere of mathematics, the following expressions are all semantically true and generally accepted by mathematicians, but are mathematically unprovable if applied to physical infinity. (Substitute at will other mathematical symbols, representing infinity, such as the aleph null, and the results are the same.)

$0+n=n$	$\infty+\infty=\infty$	$\infty+n=\infty$
$0-n=-n$	$\infty-n=\infty$	$\infty-\infty=\infty$
$0/n=0$	$\infty/0=\infty$	$\infty/\infty=\infty$
$0\mathrm{x}n=0$	$\infty\mathrm{x}0=\infty$	$\infty\mathrm{x}\infty=\infty$

'0' is a mathematical concept, as is '∞', as usually applied. Physical infinity is not. It is totally outside mathematics and therefore cannot be mathematically proven. Only logical arguments can be used.

Measurement of physical infinity

To measure something we must be able to observe it. If we can observe something, it must be measurable. Conversely, if we cannot observe something, we cannot measure it. The unit used to measure it by could be itself; e.g. 37 dots. Measurement makes it finite. Therefore motes, (The fundamental entities of the universe, described in proposition 3 of this series) which have separate existence from each other, are finite, even though they have no physical size.

37 on its own is meaningless. It must be 37 units of something. If we measure something, it must be in a unit of some kind. Unit names and sizes are purely arbitrary, and depend on what we are measuring, language, common usage etc.. The unit can be the object measured, so we could have 37 motes, but we cannot have 37 as a measurement.

There cannot be an (p)infinite number of units as the attribution of units cannot be applied to (p)infinity. Mathematical infinity differs in this respect. One can always carry out a mathematical function (e.g. +1, -1, x2, /2 etc.) on any measurement, however large or small, even mathematical infinity (e.g.∞-1). (P)infinity-1 is meaningless, just as liquidity-1 is meaningless.

A dimension can be described as of (m)infinite length, but cannot be described to be of (p)infinite length. This is an oxymoron, as maths cannot be applied to (p)infinity, and size is a mathematical attribution. It can only be described as (p)infinite. Similarly, infinitely large and infinitely small are also not permissible expressions when discussing physical infinity.

A Paradox explained.

There is an apparent paradox in the statement: Infinitely small equals zero when applied to (m)infinity. This implies that if infinity exists it does not exist. This paradox can be easily explained if we understand the difference between (m)infinity and (p)infinity properly. Whilst (m)infinitely small is acceptable, as it is a measurement and 0 is a number, (p)Infinitely small is an oxymoron, as it implies measurement. It should only be described as (p)infinite. This does not equal zero, as zero is a mathematical expression and (p)infinity is outside maths. However, it is equivalent to zero, as shown below.

The only situations that do not require units to express a value are zero and infinity. (e.g. speed can be zero or (m)infinite (a cop-out for a very large number) but it cannot be 37.) Boundaries provided by units are necessary for measurement, as one needs 'somewhere to put the end of the tape'. If something is (p)infinite (or 0) it has no boundaries in spacetime so it's size is not observable or measurable. <u>0 and (p)infinity are therefore equivalent.</u>

In a finite situation, one can always add or subtract another unit, even if that unit is of zero size. One cannot add or subtract anything from (p)infinity. $0+1=1$; $\infty+1=\infty$. <u>0 and (p)infinity are therefore not equal.</u>

It follows that (p)infinity cannot be manipulated with numbers. It is outside mathematics.

PROPOSITION 2
THAT SPACETIME IS NOT PHYSICALLY INFINITE

Spacetime

Spacetime does not physically exist. It is a construct for measurement.

Spacetime has 4 dimensions. Height, Width, Length and Time. These are all measurements, not objects or entities. Spacetime does not physically exist. It is a mathematical concept. You cannot have 37 miles. You can only have 37 miles of something or space between 2 things. Similarly you cannot have 37 units of spacetime; only things can have their spacetime measured (This envisages space itself as an object, which can have its volume and time measured). To fully describe an object in spacetime requires 4 measurements, but 1,2 and 3 dimensional entities can and do exist, as I shall show, but they cannot be fully described mathematically, due to their remaining unmeasurable dimensions being (p)infinite rather than zero.

Note the difference between (p)infinity & (p)infinite. The first refers to a state of energy; the latter to a dimension of energy, which is mathematically unmeasurable.

If we consider that the universe has no spatial boundaries, and its size is limited only as to how far finite energy, travelling at the speed of light, has reached after the Big Bang, then spatial measurements can theoretically be (m)infinite. But not (p)infinite as measurements always start at zero, which can be considered a boundary. Only the 'void' (*beware: oxymoron*) our universe 'floats' in can be (p)infinite. (The limitations of our 4 dimensional language are necessary for us to imagine we are imagining a (p)infinity! It cannot actually be imagined as there can be no image, as it is a state of energy, not an object.)

Finite energy is free to go on travelling for ever outwards (subject to time limitations described later) so this ever expanding edge to the universe does not constitute a boundary. We know that the edge is moving as the universe has been measured to be expanding and has been growing in size ever since the Big Bang. It follows that if the universe has no spatial boundaries, entities can exist in it that are spatially (p)infinite, but which cannot be mathematically measured. However such entities must always have a time dimension, due to the Big Bang creating a start boundary.

Any of the 3 spatial dimensions can be (p)infinite, (in the sense that (p)infinite = unmeasurable) but a 4 dimensionally (p)infinite entity cannot exist in this universe, as this universe has a time boundary; the Big Bang. It follows that a time infinite entity cannot exist in this finite universe due to this start boundary. (Logically, if it is finite, the universe must also have an end time (see proposition 5 of this series).)

Relativity

We are always travelling but we NEVER travel in space.

We can resolve this apparently paradoxical statement as follows:

Instantaneous travel is impossible in this universe. However small the distance it must take time. Therefore we can only travel in spacetime.

But everything in the universe is moving, all the time, from galaxy mega clusters to fundamental entities.

Stand still for 1 second. You have not traveled in space relative to the earth, but you have traveled.

In that 1 second you have traveled a few millimetres eastwards relative to the Earth's centre as it rotates; another, different distance in another, different, direction as the earth orbits the sun; yet another distance and direction as the galaxy spins etc., etc.. You will have traveled slightly further relative to the Earth but less relative to the sun if you are on a geostationary space satellite than you would if in a submarine. It is all relative.

The important thing is that over that 1 second you have not just travelled in space but in time too. You cannot travel exclusively and instantaneously in space. You can only travel in spacetime.

It is therefore equally valid to say that you cannot travel in time.

Everything in the universe, from elementary particles to galaxy clusters is travelling in spacetime relative to everything else.

There are therefore no fixed points in space. If there are no fixed points in the universe (or outside it) we cannot define co-ordinates for measuring exact, definitive distances in space or time. We can only measure spacial distance relative to other objects at a point in time.

Length, width, and height are not entities. They are measurements. You cannot have 37 metres. It must be 'of' something or between 2 objects or points.

Time

Similarly, time, the 4th dimension, is a measurement. You cannot have 37 seconds. It must be of, or during something or between 2 or more events happening or not happening. It has no substance. Despite common perception and usage, it also is not an entity. This apparent feature of time is due to our ignoring relativity.

If time is not an entity, but merely a measurement, it cannot flow. It cannot move so has no inviolate arrow of direction. Equally, it cannot be reversed. It cannot split into trousers. These are all figments of the mathematician's art. It can only be slowed or speeded up by taking measurements in relative terms.

If we cannot definitively say where we were a second ago in spacetime nor predict where we will be in the next second, there is only Now that is definite. Everything else is relative to some feature of the universe.

There is no predictable future and no definitive past. These are figments of our imagination. If we were to measure how long an event takes to happen we observe a series of Nows and then apply arbitrary units of time to determine the time dimension, relative to the appropriate environment. We cannot observe the event before it happens nor after it is over.

Just as we cannot definitively measure space as we have no co-ordinates, so we cannot definitively measure time. All such measurements must be relative.

If we measure the length dimension of a line, it is equally valid to measure it from left to right, or right to left. We get the same result.

If the time dimension is also purely a measurement it is equally valid to measure it too in either direction; from now into the future or past, or from the future or past to now, but as shown above the future or past can only be calculated, relative to a defined Now. They cannot be experienced definitively.

We can establish our relative position in time from past events (e.g. approx.13.8 billion years from the Big Bang) but we cannot definitively establish our position in spacetime in either direction, as we can have no permanent definitive co-ordinates.

A fossil is merely a relic whose spatial position relative to the earth has remained static, but relative to the rest of the universe cannot be calculated, and who's time dimension has moved through a series of Nows that can be measured relative to other Earthly parameters but who's actual spacetime position cannot be definitively ascertained, let alone that position millions of years ago, so there is no way to travel to its origin.

Taking our earlier example of standing still somewhere on Earth, if we now step to the left and then back to the right, in a period of say 1/2 a second, we only know that we are back in the same position relative to the Earth. Our new position in spacetime is completely different and unknowable as we cannot relate it to the universe as a whole.

So is our original position. That has now moved in terms of spacetime.

There is only Now. It is only our brains that recall a past and imagine a future, entirely relative to our Now here on Earth. That we go on to experience that future is only due to our experiencing a series of Nows in spacetime relative to Earth.

Our original position has moved through space and time (spacetime) in many different directions and distances in that half a second, relative to all the objects in the universe, all of which are also moving, all travelling in different directions at different speeds. so our spacetime position then is completely incalculable so can be treated as non-existent past. Just as speed is relative to the object being observed, so time, a component of that speed, is also relative to all the different objects in the universe,

We may recall being here 1/2 a second ago but that is only a function of our human brain and is entirely relative to Earth. We cannot travel back to that same spot in spacetime just as we cannot predict where we shall definitely be in spacetime in a further 1/2 second.

There is only Now. It is only our brains that recall a past and imagine a future, entirely relative to our Now here on Earth. That does not mean the past and future did not and will not exist theoretically, just that we cannot determine their position in spacetime so to us they do not materially exist.

We cannot travel instantaneously. The fastest possible speed is the speed of light. This can be measured in many ways, with purely arbitrary units, all equally valid, based on something we can definitively (relative to the Earth) measure: miles per hour; kilometres per second; lightyears per decade (If having an answer of base10 helps your equation) or hairsbreadths per century (If you want a very big number for the same reason). The key is that the unit of measure for speed of travel must contain both a spatial unit and a time unit, both based on definitive constants relative to Earth.

As an example, nothing can travel at 37 kilometres nor at 37 hours, only at 37 kilometres per hour.

We cannot travel in space alone; we cannot travel in time alone; we can only travel in spacetime. But because we cannot establish co-ordinates, there being no fixed points in the Universe to measure from, we cannot exactly know where we are in spacetime. Everything has to be relative.

Because our actual experience is all relative to Earth we ignore relativity for everyday purposes but it should not be ignored by science.

The scientific view of time seems ill defined, viewing it both as an entity and a measurement at the same time. This seems largely a myth perpetrated by maths and conflicts with science's view of the other 3 dimensions. If there is only Now, time is just a mathematical convenience, a measurement, to establish the relative position of an event, just as spatial measurements do.

There seems no valid reason why time should be treated differently to length, width and height and yet science constantly does so.

Time appeared at the start of the universe as newly finite entities created by a phase transition began to move and the time taken from point A to point B could be observed and measured (if we had been there!).

We know that everything in our universe moves, from the largest galaxies to quantum particles. Why? Because, I contend, it is all made of 'motes', spatially infinite entities, moving at the speed of light. I explain later how motes can become matter and light (see proposal 4).

Everything in this universe is moving, relative to everything else, from the largest galaxies to the smallest elementary particles, and has been ever since expansion started at the Big Bang. This enables time to be measured. Time is the measurement of the distance between 2 events, such as a particle being at point A and then at point B. Time cannot be (p)infinite, not only because of the Big Bang boundary, but also because it requires movement or change to be measurable. If something were totally stationery and unchanging (which cannot happen in this universe, where everything is made of motes travelling at the speed of light, as I shall show later and there is always the boundary of the Big Bang, where everything changed) then its time could not be measured. It would have no time dimension and so it would of necessity be (p)in-finite.

There is some argument that the Planck length and Planck time are the smallest measurable dimensions. This is a mathematical argument that is outside this hypothesis. For the sake of this work, entities are either finite or (p)infinite.

Similarly, quantum foam and the movement of elec-trons around the atom et al are measured in quanta. Quanta are not physical entities. They are measure-ments and quantum theory is purely a mathematical concept and cannot be physically demonstrated.

Anything which does not move has no time and there-fore cannot be in this universe.(P)infinite entities do not move, so cannot be said to be in this universe. However, they cannot be considered to be anywhere else either, as they have no spatial dimensions. There is no such place as 'outside this universe'. There is only a (p)infinite phase of pure energy.

The (p)infinite in any or all dimensions cannot have time as it cannot move. So a 2 dimensional entity would have a time and a length dimension, but its other 2 dimensions would be (p)infinite, so would have no time. Instantaneous action at a distance could therefore happen in these dimensions. Movement means change and (p)infinity cannot change or else it would be in a state other than (p)infinite. And vice versa; anything finite in any dimension of spacetime must move, and so must have time. Everything in the universe (except (p)infinite dimensions) moves so the universe has time and is finite.

It also follows that time cannot run backwards, as time is a measurement, not a physical entity and a particle which moved from point A to Point B and then back to the location point A, in either space or time, would actually be at a 3rd point, C, in spacetime. Time can be imagined, and mathematically calculated as going in reverse order, but cannot be physically measured (observed; see below) doing so.

This demonstrates the difference between mathematics and reality. Mathematics can model the impossible, but still cannot model the (p)infinite, despite the (p)infinite being logically real.

Einstein's proven concept that each observer has its own time scale further demonstrates this. In theory, we can only travel in time to the extent that relative time can be slowed or speeded up; all we have to do is change our speed of travel relative to what we are observing.

We can only change the speed of time by speeding up or slowing down relative to what we are observing and to do this on a sufficiently large scale to make more than a small difference is currently impossible. But we are also travelling relative to all the other objects in the universe, which are also themselves travelling, even if we are not observing them, so the speed of time is indeterminate, and there can only be Now..

It has been demonstrated by experiment that a clock at a higher altitude will be faster than one at sea level. This is because, due to the rotation of the earth, it is travelling slightly faster than the lower one.

To physically travel backwards in time, we would have to know the exact position in spacetime of every elementary particle in the universe and be able to manipulate in 4 dimensional spacetime those we wanted to affect. This is patently impossible.

An object can have one or more of its spatial dimensions in a (p)infinite phase, without affecting the finite physical size of the others. But it cannot have a (p)infinite time dimension in this universe because of the Big Bang setting a boundary. Time can only be finite. A point, in the spatial sense, is a point on the time dimension only, that has a (p)infinity of energy in the other dimensions. Again, that does not mean an infinite amount of energy, as amount means measurement. (P)infinity just 'is'. If a point has no time either, it becomes a singularity, a 0 dimensional (p)infinity, as occurs at the centre of a black hole or at the start of the Big Bang.

Warping of spacetime

I repeat, time is not an entity, it is a measurement. It is the measurement of change, the space between events; from something being at point A and point B; or from state A to state B. It follows that spacetime cannot physically warp. This is only a mathematical concept.

A point on a 2 dimensional map can be described by reference to 2 co-ordinates. The point does not have to have a physical size as the co-ordinates specify the crossing point of 2 imaginary lines. This can also be done in a 3 dimensional space with 3 co-ordinates.

Should a 4 dimensional co-ordinate system be possible, a point, having no spatial dimensions in spacetime could also be plotted. If it were not moving, no line could be plotted on the time co-ordinate, nor on any of the spatial co-ordinates. This point would be a (p)infinite singularity. If it moved in a straight line in the spatial dimensions, it would be a finite point and the plot would describe a simple curve in space time, as it also moved through time, while if it moved not in a straight line, or at irregular speed, the plot would be a complex curve. This enables the logical proposition of the 1 dimensional mote as the fundamental entity in proposition 3.

The bending of light rays by gravity is a physical phenomenon, due to motes, of which the rays are made, having mass, and therefore susceptible to gravity, but cannot be explained by the physical warping of spacetime. It can however be explained by the mathematical concept of the warping of spacetime.

This suggests that other formulae may be describing events or describing things that are not physically correct, but are considered as proven by the scientific method.

As we have said, measurements of the (p)infinite cannot be made, as it is unobservable, so 0 dimensional (p)infinity can have no spacetime.

PROPOSITION 3

THAT, IN ANY UNIVERSE, ENERGY MUST BE A FUNDAMENTAL ENTITY

The current view

Quantum theory is the currently prevailing physical theory, However, to have 17 (or more, some say) elementary particles with multiple flavours and 4 forces, plus anti-matter, ready and waiting to inflate together at the Big Bang seems a little unlikely. Where did they all come from? How did they all squeeze into a singularity? These particles may be elementary, in the sense they are what elements are made of, but that many cannot be fundamental. We have seen that only one thing can exist in a (p)infinity. That one thing must have transferred to this universe at a fundamental level for the Big Bang to make any sense.

The Big Bang theory is the currently prevailing cosmological model for the universe. It has been shown that the Big Bang happened about 13.8 billion years ago, followed by rapid inflation, which cannot be explained by general relativity, but is consistent with the acceleration to be expected from standstill to the speed of light if energy were suddenly introduced at very high density, slowing relative time.

(p)infinity

We know that energy cannot be created or destroyed. It could not have suddenly been created at the Big Bang. It must have been pre-existant. It must therefore be, other than in this universe, (p)infinite and must have existed in a (p)infinite state before the Big Bang. Energy would have been the only thing in that (p)infinity. A (p)infinity cannot hold more than one thing, or it would not be (p)infinite for either of them. We have no words to fully describe that state, other than to suggest it is a phase of being. It is not an area or a volume, or anything we can imagine, as all such have boundaries. (in this sense (p)infinity cannot 'hold' anything, but oxymorons are needed when talking of (p)infinity)

The Big Bang explained

But something happened in that (p)infinity. Something changed. A phase transition. I have no explanation as to why. Perhaps it is a random event and there are an untold number of universes similarly created. The (p)infinite phase would not be affected in any way. However much finite energy appeared in the new universe, it would not affect the (p)infinite. As a (p)infinity has no numbers, however much energy is syphoned off for each universe, the (p)infinity remains (p)infinite. (To put it mathematically $\infty - n = \infty$)

Energy, which had been locked unmoving in the (p)infinity now had room in which to move. As soon as it started to move from point A to point B, the measurement of time was possible (even though we were not there to measure it ourselves.). We can describe this as time starting. Something must move, or events happen, for time to be measurable.

(P)infinite energy could not move, as it has no boundaries in which to move. Now as it moved, it created boundaries, i.e. starting to move (time boundary), where it had moved from and where it had moved to (spatial boundaries), so it had to become finite.

Finite entities are measurable, but (p)infinite energy has no dimensions to measure, and, in any case, it had to have units to be measurable. So it became an untold number of moving entities, or units. They would have a measurable time dimension because they move and have energy to make them move but nothing else. No spatial dimensions at all. They had no substance to make them particles. Points of energy.

Without time, fundamental energy cannot be said to travel, as this requires time and spatial dimensions to be measurable. It cannot even be said to be in a (p)infinite number of places at once, as this implies spatial dimensions. It just is.

However, once time is in the universe, the concept of the mote (my name for these points of energy- see later for explanation) of energy, travelling at the speed of light, becomes possible.

Conversely, there can never be a (p)infinity of elementary particles as they occupy all 4 dimensions, and any dimension is measurable. At CERN they claim to have seen these particles translated by mathematical calculations to display onto a screen. If they can be observed, they must be measurable, so must be finite. At the Big Bang there must have been a measurable number of particles if current theories are correct. As we theorise that these particles have physical size, they could not have occupied zero space. It follows that the Big Bang could not have started from a single point in spacetime if elementary particles are the basis of the universe.

The creation of an energetic entity

Finite Energy has to perform work, The only work it could do at the Big Bang was moving. From stationary, energy entities would have accelerated giving rise to the inflation of the Big Bang. I am unclear (I have been unable to find a reference) as to whether the calculation of 13.8 billion years for the age of the universe factors in the relatively slower speed of time when the universe was at its densest, but this would speed up as inflation increased and lowered the density, and hence the gravity. This could result in the units of energy apparently exceeding the speed of light when calculated at today's speed of light, if this adjustment is not made, but they could not have done so when calculated at the speed of light at that time. In other words, the same speed as today, but from a relatively speaking different point of observation.

The points of energy would accelerate rapidly up to the "Speed of Light". I inserted the parentheses, because light did not exist yet, as we shall see. Mass is resistance to acceleration, so these points would develop mass. The acceleration is caused by the energy contained in the point, which must be a fixed amount as these points are the fundamental building blocks of the universe, so must be all the same. By the same logic they must all have the same mass. A fixed amount of energy can only overcome a certain amount of mass, so the acceleration is balanced when this state is reached. The speed of light is created.

Once most of these motes (as I call them – see later) were at the speed of light, inflation would slow, but the universe would continue to expand as 'pioneer' motes continued to move outwards. Relative time would also speed up as the density of the early universe reduced rapidly. Because motes have mass, they are self attractive. They create a force between them we call gravity, so the lower the density, the less the gravity, and the relatively faster the time. The force of gravity must be lower than the force of energy causing the motes to move, allowing the Big Bang to happen.

Gravity also means that the motes would roil and swirl, further slowing inflation, as individual motes within the mass were attracted to each other, but did not join but 'sling-shotted' around each other, again and again, in the same way that space ships will sling-shot around a planet by using its gravity and that of adjacent planets and the sun.

This slingshot affect repeats again and again at all scales throughout the universe, down to sub-atomic levels, and up to Galaxy mega-clusters, enabling a GUT to be envisaged.

Occam's razor

This hypothesis suggests just one (p)infinite entity, energy, started it all by a phase transition and no forces pre-existed the Big Bang but only came into existence when energy started moving. Compare this to the 17 particles and 4 forces of the standard model. Apply Occam's razor and decide for yourself which explanation is more likely!

We may hypothesize as follows, but a theory using the scientific method I believe to be unlikely, as the mote, although finite in time and mass, is supposed as spatially infinite, so cannot be observed or measured.

And therefore calculations and formulae are impossible.

The Fundamental Particle -the Mote

Mathematically provable theory already exists for the observable world above the level of elementary particles, (or particles from which elements are made). Elementary particles have been mathematically calculated as existing but have not been directly observed, other than as simulations on screens made of atoms many times the size of the particle being observed, derived from complex mathematical calculations. That is not direct observation. But, if they exist, what are these made of? They have physical size so are theoretically divisible.

They are measurable (Measurable and observable are semantically the same. I prefer 'measurable' as it better describes what is meant, and covers items which technology cannot yet directly observe), so they must be made of something. It was once thought we could not split the atom, Can we split the quark? In theory, yes! It has a physical size,(or else it could not be measured or observed) so half a quark is conceptually possible.

There must be a fundamental particle, or entity, common to them all. It seems improbable that multiple particles, together with multiple forces, all arose from nothing, separately and simultaneously, at the Big Bang. Especially so as the Big Bang is supposed as arising from a singularity (i.e. (m)infinitely small volume) yet each individual particle has some volume, (even if it has no mass under current theory!). How could an untold number of particles with some volume, however minute, fit into a point with no volume? It must be magic, not maths!! (See proposition 5 of this series to see how it can be done with motes in a black hole)

At the start of everything it seems logical that only one thing would be created first (I believe in (p)infinity, so there was no 'start' or 'first', it just 'became non-infinite' but you have to express it somehow.).

 So, let us consider what a fundamental entity should be like.

It should :

- Be Indivisibly small; That means it must be (m)infinitely small; **a point,** a spatial (m)infinity. Anything larger (i.e. having any measurable or theoretical volume) could theoretically be divided, and if it can be divided it is not fundamental.

- Have mass; If $E=mc^2$, (and this is one mathematical formula I do believe to be true.) m cannot $= 0$ or else $E= 0$ also, which is patently absurd. Anything with zero energy and zero mass could not be said to exist in any real sense (in this universe at least. (P)infinity is another matter!) as it could have no other properties either. Despite this scientists claim zero mass particles do exist, and even whizz around at the speed of light, despite, of necessity, also having zero energy. (photons! See proposition 5 of this series for debunking). They obviously don't agree with Einstein, or else it is one of those damn mathematical constructs again (i.e. cheats or fudges)!

- Have energy; (obviously from above)

- Be unique i.e. be the only thing existent at that level; If other things exist then it is not fundamental.

- Be permanent. It cannot bubble in and out of existence. It must always and for ever exist, and, by implication, not just in the time this universe exists. As the (p)infinite does not have time, outside this universe it must be in a (p)infinite state.

That is about it. Any other properties would be extraneous.

But something of (p)infinite size cannot really be described 'to have' anything. This implies it contains something smaller and we have seen that a (p)infinity can only be of one thing. Nor can it be called a particle; if it is indivisible, it can have no physical size and only be a point; at best an entity. We should therefore say it has no volume, but is energy, and is mass and is time. Of these, mass and time are unique to this universe (or any similar universes which may co-exist.).

I shall call it a **MOTE,** from an anagram of the acronym of Only Energy, Mass and Time.

Speed is distance over time. Can it be a coincidence that the most famous formula of all, $E=mc^2$, is comprised solely of these same elements, energy mass and time?

Substituting in this same formula we get $c=\sqrt{\dfrac{E}{m}}$. As previously described, the speed of light was achieved when the energy of the mote was balanced by its mass.

Properties of the Mote

There are other things that logically follow from the above:

- Energy cannot be something contained within the Mote, or fundamental point, as it is indivisibly small.(An oxymoron, as size cannot be applied to (p)infinity, but just saying it is

(p)infinite does not convey a lack of volume-
due to our normal concept of (m)infinity being
extremely large or extremely small.)

- If energy were emitted from the mote as a
  wave, energy would be lost so the mote, the
  basic unit of energy, would be no more. The
  wave would also give the mote volume.

- It cannot be a force field if the Mote is a point,
  as this would give it volume as well.

- Simply pulsing or vibrating (as in string theory) cannot explain it as the Mote has no
  volume so any pulse or vibration could not
  move more or less than the size of the Mote,
  which is zero.

- The mote cannot rotate, as that would imply
  volume to rotate.

Energy must therefore be a property of the Mote. The
mote is energy.

The only possible action it can have is moving. Below I discuss how it can be in all places at once, or
none at all, at (p)infinity, but the presence of time in
the universe means it can only be in one place at one
instant in time, (actually only true of its mass which is
its only finite property) and consequently in a measurable path in any sequence of time. This movement
must be in a straight line, (when not affected by an ex-

ternal force) as any deviation from a straight line would imply that the Mote had asymmetric volume.

As all Motes are fundamental, and therefore identical, they must all travel at the same speed dictated by their energy; the speed of light

If this is fundamental, it confirms that nothing can travel faster, as anything made of motes (which everything is, as we shall see.) cannot travel faster than a single mote.

Why would a mote choose a particular direction to travel in? It theoretically does not. It can travel in every direction at once. Being a one dimensional (p)infinity (it only has time) it can be everywhere, but not at once in this universe. Its mass property limits it to one place at any particular point in time.

But it can be instantaneously entangled with other particles as it has no spatial dimensions, so, despite being a point in the time dimension, it is everywhere (and/or nowhere!) in the spatial dimensions. (Oxymorons again)

I find it easier to think of a mote as a moving point in spacetime with mass but no volume, rather than as a spatially (p)infinite energetic entity, although it is both as they are the same. However, one cannot imagine a point successfully, as we 4 dimensional creatures will always think of something physical being there.

To imagine it is impossible as there can be no image as an image must have spatial dimensions.

Detecting the Mote

A single mote, being a point, is undetectable by normal physical means, but if they exist in vast numbers, we might be able to detect them. The only energetic entity not considered a particle or a force is heat, so we can conjecture that motes in large numbers can be measured as heat.

Heat can be considered to be the detection of motes en masse, whether by the body or by instrumentation, as the energy of an individual mote would be so small as to be undetectable and any detection equipment would also be made of innumerable motes and could not be at absolute zero as explained in proposition 1 of this series.

When things get hotter or colder, the heat energy does not lose or gain strength. It is merely the concentration of motes as they flow from one place to another, trying to level their concentration out due to gravitational effects. It is the concentration of motes that determines the measurable temperature. The more motes in a given volume, the hotter it gets.

Absolute zero means no moving motes. If it were reached, everything at that temperature would destruct, or phase change, back to (p)infinity as the motes stopped moving, meaning time had also stopped for them.

They can change direction by the slingshot effect of their own gravity as they approach, but do not quite meet, each other, and we can suppose that the denser they are en masse, the more they will change direction until that density lessens. Thus spacetime contains a roiling cloud of motes, perhaps explaining the phenomenon of quantum foam observed by physicists

Black body radiation has been shown to move at the speed of light, confirming it is motes that are moving. The overall heat energy of the universe as a whole must be relatively constant, explaining the CMB, although some motes will constantly be returning to (p)infinity in black holes and the like.

It seems to me that mass and time, being created by the movement of motes, are a consequence of energy, rather than being fundamental properties of matter.

The time between a mote being at point A and point B can be measured (Theoretically at least). As the mote is a finite entity, and does not exist in (p)infinity, its time can only be measured in this finite universe.

Motes are pure energy but they must have mass, however small, as they accelerated initially at the Big Bang. As the motes travel, their speed, after the acceleration phase, must be dictated by the energy available to move their mass. The speed of light is established by the energy of the mote being balanced by its mass. It is the same speed in all subsequent matter and light as everything is made of motes, as we shall see, and they are all identical, being the fundamental entity in the universe.

If energy moves mass at the speed of light over time, neither mass nor time would exist if energy did not do work, but energy is not dependant on mass or time existing. That is, if energy were static and motes did not move, time and mass could not be said to exist. This is the case at (p)infinity where energy just 'is'.

If a mote were not travelling, it could not have mass, as it would not be being accelerated and mass would not be necessary to mediate that acceleration. Time could not exist as it could not be measured if everything were always the same and the energy would be (p)infinite.

The efforts at CERN and other particle accelerators cannot ever discover motes however small they break matter into, because, if motes have zero volume, they cannot display a mark upon a screen, because everything that screen contains will be 'bigger' than motes by many orders of magnitude (although the atoms of the screen are made of combined motes (c-motes; see proposition 4 of this series), and ultimately of motes at a fundamental level). A point entity (spatial infinity) cannot be detected by finite particles.

(Beware the oxymoron above. 'Bigger' should not be associated with a (p)infinite entity, which has no size, but there is no other way to describe it.)

An individual mote could not be detected as its contribution to the background heat that pervades everywhere (for motes are everywhere in quantity, even the coldest environments) would be vanishingly insignificant as would be its mass.

In any event, it is moving at the speed of light and cannot be slowed or stopped during observation, as this would mean it was destroyed back to (p)infinity. Observation of its physical presence is also impossible as it has zero volume. (NB 'Destroyed',above, does not mean physical destruction. This is impossible. A stationary mote phase changes to being (p)infinite energy.)

However, p-motes (Again, see proposition 4 of this series) probably will be observable when they smash particles together, as the motes created recombine close to the event. They should look for EMR of any frequency, at less than photon size. The photon is usually the smallest EMR particle observable, but this is a 4 dimensional bundle of 2 dimensional p-motes (see later).

It seems ironic that such huge machines have been built to detect particles larger than those fundamental entities that our own bodies already detect: motes as heat, and p-motes as light and c-motes as the matter of which they (and we) are built. (See next proposition)

Existing particle physics compared

Lastly in this proposition, let us consider the existing group of items currently described as elementary particles: electrons, quarks, fermions, barions, bosons, et al.. and their further 'flavours' and antimatter as well. To have that many differentiated particles just does not seem 'fundamental' i.e. they all appeared, ready made, at the Big Bang.

Another problem is that the Big Bang is envisaged as a singularity. How can finite particles appear in their near (m)infinite number, enough to create a universe, at or shortly after a singularity which has no dimensions for them to appear in? Did each particle start (m)infinitely small and miraculously increase in size?

One would expect only one thing to appear first at the beginning of the universe, and if it was a singularity that thing should be capable of (p)infinity. However, as far as I am aware, science at its present stage cannot describe anything smaller than the elementary particles. Elementary particles should therefore be described as the smallest *measurable* particles, and nothing more.

'Elementary' particles in the sense of particles that combine to make elements will suffice as a name. I believe that they are all, in fact, made of spatially (p)infinite motes at the fundamental level. (See also proposition 4 of this series)

PROPOSITION 4
THAT THE FABRIC OF THE UNIVERSE IS COMPOSED OF THE MOTE AND GRAVITY ALONE

Mass

A current definition of mass is '*a property of a physical body which determines the strength of its mutual gravitational attraction to other bodies, its resistance to being accelerated by a force,*' (Wikipedia Nov 2015)

The mote described in proposition 3 of this series acquired mass as its energy required it to move after the phase change from (p)infinity at the Big Bang. It accelerated from not moving and continued to accelerate, acquiring more mass as it did so.

The mote no longer accelerates, but travels at a constant speed. It first accelerated at the beginning of time at the Big Bang, until its mass balanced further acceleration dictated by its energy. This was at the speed of light, and motes have travelled at this constant speed ever since. Its mass continues to balance further acceleration.

In other words, the energy a mote contains will only push its mass at the speed of light. No more and no less. (Or, to put it another way, its mass is created by the movement of its energy, measured as resistance to that movement.) Because this universe has spacetime, spatial dimensions give us the ability to measure the extent of that movement and time allows that the speed is measurable.

Energy

So how can our universe be formed from spatially infinite moving particles? (The mote is not actually a particle, as it is spatially (p)infinite. It is a 1 dimensional (time) entity, but particle is easier to envisage and to say than a '1 dimensional (time) entity'.)

As energy is spatially (p)infinite, motes cannot be slowed or stopped, as there is spatially nothing to slow or stop, nor can they collide, as there is spatially nothing to hit. They are equivalent to zero in size, but they exist and have mass and time. Hence (p)infinity is equivalent to 0, but does not equal 0.

However, motes are attracted to each other. Their mass gives them gravity, however small. At this stage gravity is the only force in the universe. It does not exist without the universe as nothing moves in (p)infinity so cannot have mass. And, because their mass is minimal, (one might say infinitesimal) so is their gravity, so they have to be adjacent to overcome their tendency to continue on their own path. One cannot say touching, as there is nothing to touch. If there is the slightest gap, in space or time, they may be diverted by gravity but not join. Time must be included, as they cannot speed up or slow down to catch each other.

If the above is the case, a mote is nothing but a packet of energy, and everything in the universe arises from it. The terms Mote and energy become interchangeable.

So energy is the source of everything in the universe as we shall see.

Gravity

Gravity can be considered as the attraction of one mote to another, and is a property of the mote. It is not energetically adjustable. Simply, the more motes present in a given volume, the higher the mass and therefore the higher the gravity.

Motes travelling on the same path, one behind the other, in space or time, will always maintain a constant distance apart as their speed can only be identical. This is logical, as they each attract each other with the same amount of force so, as they travel, the forces are always balanced by their mass and speed so the one in front cannot drag the one behind towards it, speeding it up, and vice versa. They will not approach each other when on the same path.

The universe is filled with Motes all travelling at the speed of light in every direction. It is therefore logical that the universe is expanding as all motes move in random directions, as gravity pulls them one way and another, so many of those Motes will be travelling away from the edges of the universe. As it expands the relative gravity throughout the universe will reduce, as it becomes less dense. This will lead to hyperbolic expansion as the motes nearer the edges are less diverted by the gravity emanating from the more dense centre and so proceed on a straighter and straighter course outward at a progressively faster rate, until more and more are travelling at the speed of light away from the universe.

As the Motes expanded at the Big Bang into the new universe, they were changing state from (p)infinite. Why should the potential universe contain other forces, waiting to act on motes after a phase transition? There is no logical reason for it to do so or even for a potential universe to be there and waiting.

Therefore, forces must also be assumed to be a property of motes, rather than some separate manifestation, just as motes' movement created mass and time. The creation of motes, together with mass and time, created gravity. I shall demonstrate later that, without the current mathematical theories that create them, the fundamental forces other than gravity envisaged by current mathematical theory are not necessary for a GUT to be formulated.

Gravity cannot occur without mass. In (p)infinity there can be no gravity as nothing moves, so there can be no mass.

The mote is purely moving energy which creates mass; time and gravity arise from that fact. It could be said that when 2 masses approach each other, gravity occurs, but does not exist without this juxtaposition. But we have said that mass is a consequence of energy accelerating. Gravity must also be a consequence of mass.

However, if gravitational effects are constant it would not seem logical that expansion of the universe is speeding up, as has recently been observed, unless one takes (p)infinity into account. Then it seems perfectly reasonable.

The only expansion that starts at (p)infinity, as the universe did at the Big Bang, follows a hyperbolic curve. All other types of expansion have a start boundary in spacetime.

We can therefore conjecture that the universe is expanding on a hyperbolic curve which will eventually again return to (p)infinity as all hyperbolic curves do.

Expansion must be hyperbolically slowing or increasing, dependant on where we are on the curve.

The Wikipedia (Oct 2015) definition that mass is '*a property of a physical body which determines the strength of its mutual gravitational attraction to other bodies*' assumes that gravity is a separate force extant in space since the beginning of time for some reason unknown to current science. Suppose that it is, in fact, a manifestation of energy contained within the motes. i.e. motes are fundamentally attractive to each other.

Similarly, we can infer that the other forces envisaged by mathematics are also created due to an integral property of the mote. Motes are energy, so natural forces are manifestations of energy contained within the fabric of the universe, not some separate entity.

This implies that, at a fundamental level, there is only one force, the attraction of motes for each other.

The separate identities of the other forces only become mathematically apparent as the elementary particles appear. This is why physicists (mathematicians all) have had to invent gluons, because they cannot conceive matter arising from pure energy, as to do so it has to include (p)infinity where maths cannot go.

Light

If motes are travelling adjacent to each other on parallel paths as they will when packed together immediately after a bursting event (see later for definition) or If their paths are travelling in the same general direction, but are approaching each other and nearly parallel, when they are close enough, their gravity will pull them together until they are adjacent. Although individually they have no spatial dimensions, their mass constitutes an entity. They cannot cross each others paths as they cannot change speed so will then both be slightly diverted and continue on a new joint path together. They can then be considered as one unit. Their speed cannot change, and they cannot individually change course. This type of joined mote is 2 dimensional (Width and time). I call this type of joint mote a p-mote, after **P**arallel.

More than 2 motes could conceivably be adjacent and could join. In this case, the p-mote will be 2 or 3 dimensional (time, and width, and possibly height), but it can never be 4 dimensional, as motes cannot catch up with each other, so length must always be (p)infinite.

It should be noted that motes have zero spatial dimensions, but cannot fill exactly the same space, otherwise there would be 2 types of mote due to their combined mass, and they would not be fundamental, so although this apparently gives rise to 0+0=more than 0, we have said (p)infinity is equivalent to 0, not equal to 0. In any event, 'the same space' is meaningless if spatial dimensions are (p)infinite.

Beams of such joint motes are what our retina interprets as light, or at other frequencies can be detected as part of the EMR spectrum (See below for redefinition of EMR). P-motes themselves hold no information. It is the gaps that provide frequency information. Immediately after the Big Bang the new expanding universe was filled with accelerating motes, moving away from a point, packed together, so many of them would be expected to join as p-motes. We cannot say the universe was filled with light, as light is merely an interpretation living beings place on patterns of p-motes at certain frequencies, travelling in beams, and we were not around then. Also, all, or most frequencies of EMR would have been present, including electricity, x-rays, gamma rays, etc..

Light waves cannot exist, as waves must travel in something, and decay too rapidly whereas beams of light can exist for millions of years (in fact indefinitely) through space, so that we are able to see distant stars. If motes are the fundamental entity, and are energy, any such waves could not be made of anything else as there is nothing else for them to be made of, either material or energetic. Also, in bright light (A high concentration of p-mote beams) waves would interfere with each other, creating 'chop' to use a nautical equivalent, thereby destroying our ability to interpret it successfully and see detail; quite the reverse of our actual experience of bright light.

EMR

Electromagnetic Radiation is classically described as consisting of electromagnetic waves. As seen above, light does not have waves, so it can be deduced that nothing on the electromagnetic spectrum has waves either. If the waves travel at the speed of light, then whatever the waves consist of must be travelling faster than the speed of light, as the wave form travels further than the covered distance of the beam. This is impossible as nothing can travel faster than a mote, as we have shown above. Such oscillations are purely mathematically generated to describe the (p)mote/gap/(p)mote structure of electromagnetic beams described above. I shall continue to refer to EMR, but under this redefinition.

Matter

Alternatively to combining into p-motes, motes will also meet as follows:

If their paths are at a steeper angle, or opposing, as they approach each other their paths will be diverted by their mutual gravity. Then instead of following the same path when adjacent as for p-motes, some paths will lead to a sling shot change of direction only, but some will cause the motes to spin around each other, in orbit of each other. They will remain so unless disturbed (see Bursting Events). I call such joint motes **c**-motes, for **C**ombined motes. The speed of spin will remain at the speed of light, but their speed of travel as a joint entity can slow or even stop (relatively). Spinning at the speed of light in such a small space would make them, for all intents and purposes, solid and 4 dimensional.

It should be noted that motes have such small mass that they will only join by gravity if they are actually adjacent. At greater distances, they will sling shot around each other in the same way that a spaceship can be guided around a planet, or light beams bend around a massive cosmic object. Consequently, in a large volume of motes, all travelling at the speed of light, only a few will join to form c-motes. Most will simply roil about, pulled hither and thither by the constantly changing gravity acting on them.

This type of joined mote is 4 dimensional, because of the differing angles at which they meet. There is a notional 3 dimensional possibility, but this is likely to be very rare, when motes meet directly in line with each other.

The c-mote is the basis of matter. At the size of only 2 joined motes it is likely to be undetectable with current equipment, as it is only just above (p)infinite in size, but as more motes join the cluster, the quantum particles may be formed (if they exist; see Proposition 5 in this series) The larger the cluster grows, the greater its gravity, and the more likely it becomes that additional motes will be attracted.

From that point, matter as we know it is detectable. C-motes can be said to be the basic quantum particles.

Heat

As seen before, motes which fail to join can still be detected, if in sufficient quantities, as heat. The more motes present, the higher the temperature.

I doubt it possible to measure a single mote as it will be travelling too fast and there will be too many background motes, even at very low temperatures.

Bursting Events

Because individual motes have so very little gravity binding them into p-motes and c-motes, they are very easily separated at what I call Bursting Events. These occur when further energy is applied to a material, such as pressure, friction, some chemical reactions and the application of energy directly in the form of heat (more motes) .

Bursting events include nuclear fusion & fission; friction; electric arcing; etc.. (& black holes & especially the Big Bang). They can even be as simple as striking a match or polishing a surface or creating pressure in

a bicycle pump. With the addition of fuel to aid the reaction, also fire. P-motes in light can also be used to
release motes as heat in laser cutters. Such events as
lightning or explosions are more obvious bursting
events. Radioactive materials are where bursting
events take place spontaneously and continuously.

C-motes (in atoms or their elementary particles) or p-
motes (in photons) are forced apart, individual motes
are released and fly off in a straight line at the speed
of light. Anywhere heat is given off is a bursting
event, where c- and /or p-motes are being returned to
pure motes. Black box radiation is purely motes travelling away from a bursting event at the speed of
light.

Typically, these events are accompanied by the observation of heat and light (motes and p-motes in quantity). Light and elementary particles are created as p-
motes and c-motes which easily recombine from
motes which are closest together at or immediately
following the bursting event.

However, each mote from any particle has to be released as a discrete instance and therefore at a separate point in time. This forces regular gaps into the released beam. In the case of p-mote beams this creates
EMR frequencies.

The more energetic the bursting event, the higher the
frequency of release, and the closer together the

beams resulting in stronger (brighter) EMR. Proximity of the beams of motes enables more p-motes to be created. The p-motes will be closer together in the beam the more energetic the bursting event, as the increased energy will result in more frequent release in the burst, resulting in a shorter 'wavelength' being detected.

An example of a bursting event

An example of a bursting event would be making fire by rubbing 2 sticks together. As the sticks are initially rubbed, motes are burst from the atoms of the wood by friction. These fly off in beams from the spot where they were burst free at the speed of light and can be detected as heat. Because the bursting is initially not very vigorous relatively few motes are released, so only a slight heat is felt, and as the beams diverge this quickly dissipates with distance. As the rubbing becomes more vigorous, many more motes are released and are closer together, and more heat is felt. At this stage, some will join as p-motes and will fly off in beams of low frequency light. The wood will appear to glow red. More rubbing and with the addition of oxygen as fuel, the p-motes will be seen as flames, of a brighter colour, as the extra density of motes generated create more p-motes and, in turn, shorter wavelengths as the p-motes are closer together in the beams. The flames will appear to flicker as the beams are diverted multiple times by the ever changing gravity of adjacent beams. From the bursting of each individual atom, the beams will be given off on slightly diverging paths.

As the beams are so close together there are sufficient to make photons appear to be further from the wood but as they diverge, the photon, which is a bundle of p-motes, can only be seen for a fixed distance before it decays into individual p-motes which cannot be observed. The beams of p-motes will always diverge until they are in insufficient quantity to create discernable photons, so the flame has a limited length, although heat can still be felt for some distance due to the detection of motes in quantity and (p)motes no longer forming photons.

Due to the oxygen as fuel, the reaction becomes self-sustaining and the wood burns without further friction. It will continue until all the atoms in the wood have burst apart or have changed form to ash as new atoms are formed by the formation of c-motes within the ongoing bursting events.

Charcoal is formed when the bursting events are not vigorous enough to break apart carbon atoms in the wood, but can break apart other elements, given off as motes and detected as heat and a red glow of low frequency p-motes.

PROPOSITION 5
THAT ALL CONSEQUENT PHENOMENA ARISE FROM THIS VIEW OF THE FABRIC OF THE UNIVERSE

The above explanations for a universe made entirely from Motes lead to logical and simple explanations of other scientific phenomena, where paradoxes and other problems often arise when explained mathematically by the scientific method, as is current practice.

Photons

I refer in Proposition 4 to p-motes rather than photons. The 4 dimensional photon is a strangely imagined creature. It is said to have 0 mass and therefore 0 energy, which makes no sense as it travels at the speed of light. It is 4 dimensional, so would interfere with itself in bright light when many are present.

My explanation of the photon is that it is a section of a bundle of 2 dimensional p-mote beams. A single p-mote beam would be undetectable, as it is made of p-motes which have only 1 (or possibly 2) spatial dimension (+ time), travels very fast, and cannot be slowed or stopped. We can see through the beams as they have no physical matter in the length, and only having mass in the width. The length of the gaps in the beams reaching our eye, or other detector, gives information which can be interpreted as different types of EMR. A p-mote beam on its own would be too small to be detectable, as all detectors are made of much larger atoms, which have gaps between their constituent particles through which the beam could simply pass, so when we detect EMR it must be bundles of beams we are detecting. A photon must be at least 2 p-motes long to hold any information, but this is again likely to be undetectable. I suggest that what is observed as a photon is in fact probably hundreds of parallel beams, at least tens, if not hundreds, of p-motes long, so small are these entities. This would be the minimum required to leave a mark on a screen, or our retina, each made of relatively massive atoms.

An individual p-mote would hold no information. It is the interpretation of the gaps that counts. Our brains have learnt to interpret certain frequencies to be the colours. Colours, per se, do not exist except in our heads. All there is are p-mote beams of different frequencies flying around at the speed of light and our brains interpret those of certain frequencies that enter our eye as different colours.

There is no way to determine that what I think of as pink is the same 'colour' as what you describe as pink, only that the light has the same frequency. Colour blindness would suggest that we perhaps do not all interpret the same frequency as the same hue.

An individual p-mote beam would be too small to be detectable, being only one order of magnitude above zero. Our eye certainly could not detect it, being made of atoms many orders of magnitude larger than a p-mote, so an individual beam could not affect it.

A photon's colour will be dependent on the size of the gap between p-motes in the length off the beam to give its frequency. In particular, a white photon must contain many beams of each colour frequency to make up white. Physicists may believe they are generating an individual particle, when they claim to generate individual photons in experiments, but these must be made of many, many p-motes to be detectable. A single p-mote is so small (and carries no information) that it would be undetectable. <u>A p-mote is an entity of mass, width and time only. i.e. it is 2 dimensional. It is not a particle</u>. It is not until they are gathered into sufficient numbers to form a photon that they become a 4 dimensional particle.

(Note that 'small' above is yet another oxymoron as the motes a p-mote is made of have no spatial dimensions)

It is claimed that a photon has zero mass. This is impossible if we are to believe $E=mc2$, as it would have no energy either, and its speed would be zero, which is clearly not the case. It has a vanishingly small mass, but a mass nevertheless. It may be impossible to measure that mass, as it cannot be stopped or slowed from the speed of light, but it is there. Stars can be shown to be massive, and they lose that mass as they burn. Burning is just a continuous bursting event where only heat and light are given off. If the star loses mass, then that mass must be contained in the heat and light.

Non-acceleration of light

When light is generated, it does not need to accelerate from zero speed. It is immediately at the speed of light. To explain this, take as an example switching on an electric light.

As the light is switched on, the electric current, (which is also EMR so is made of p-motes,) hits the tungsten filament and creates a bursting event . C-motes in the filament and p-motes in the current, break apart and motes are given off in vast quantities. Each mote must be given off at a discrete moment in time, so there are gaps between them, however small. Because the beams of motes are close together close to the bursting event, new p-motes are created in beams and subsequently the beams join as photons. Some motes do not recombine and can be detected as heat. The vast numbers of p-mote beams, far more than the affected atoms or their particles would account for under current theories, result in a bright light.

A tungsten filament bulb has a limited life because this erosion of the c-motes in the filament results in it being weakened over time. That this takes several hundred hours is testament to how many motes are involved in the construction of atoms (See proposition 6 in this series)·

The light does not need to accelerate, as it is made of p-mote beams, the constituent motes of which have been travelling at the speed of light since they were created at the Big Bang. They were previously, before the bursting event, in the form of p- motes (motes travelling in parallel <u>at the speed of light</u>) in the electric current, and c-motes (Multiple motes circling each other <u>at the speed of light</u>) in the tungsten atoms. As, in both instances, the constituent motes are already travelling at the speed of light, they do not need to accelerate.

Fluorescent light and LED

Fluorescent light and LEDs give off very little heat. This is because they are not created by such vigorous bursting events.

We can speculate that as the current meets the gas or LED, only the atoms in that burst and the motes released infill the gaps between the p-motes in the electric current so that they change frequency to that of light, in the same way as previously described for how we see colour. The light given off is made of the same p-motes as the electric current entering plus some from the medium in the tube. Only the gaps between p-motes (wavelength) in the beam have been changed. Electricity and coloured light are both EMR, so this is entirely plausible. As there are few or no free motes created, little energy is given off as heat, and the process is therefore more efficient, less electricity is required for similar light output.

The twin slit experiment

The very basis of the twin slit experiment must be suspect. Physicists believe the photon is an elementary particle within the nucleus of atoms. Physicists also claim that it is single photons making the marks which combine to make the interference bands in the experiment. The nucleus is 1/10,000 the size of the atom overall due to the surrounding electrons. How can a particle that small make a visible mark on a screen, itself made of atoms, when we cannot even see individual atoms?

However, let us ignore that for the purpose of this discussion.

If the photon contains many, many p-mote beams then the twin slit experiment can be explained with 2 Factors contributing to the result:

1) As the photon or beam is generated, it is impossible to direct it accurately. The smallest hole that could be made in a screen at the source of generation, to direct the photon or beam, must be at least an atom in size. A p-mote is many orders of magnitude smaller than an atom. Therefore the photons emerging through the hole will be composed of diverging p-mote beams as it is generated. Even though the main bulk are aimed at one slit, some p-motes must reach both slits, even if there are insufficient to make discernable photons. The fact that even in those experiments that claim individual photons are being generated, the results from firing many photons through the slots form vertical lines shows the inability to direct the photons accurately.

If accuracy were achieved, the result would be a row of small dashes, the height of a photon, not lines, as each photon hit exactly the same spot in the height of the screen.

2) The screens in which the slits are made are at least several atoms thick. Although the p-mote beams will travel approximately in straight lines they will be diverted by gravity in close proximity to air molecules and other structures in the laboratory, (hence we can see a bright light beam from the side. We are seeing beams diverted from the main beam by close contact with air molecules,

having a sling shot effect due to the gravity of
those air molecules)and in the screen they will be
diverted by gravity from the molecules of which
the screen is made. Those nearest the left edge
will be diverted left and those nearest the right
will be diverted right in exactly the same way that
a space ship will be slingshotted around a planet
by its gravity. Some beams through the centre of
the slits will not be diverted at all, but the beam
has been split to such an extent that those passing
straight through will be insufficient to form discernable photons, especially as very few beams
will be travelling in a directly straight line due to
the dispersion after travelling through the multiple
thicknesses of atoms. Those rejoining from the 2
sides of the adjacent slits will recombine into
photons in the well known interference pattern.
The result from the 2 slits will be vertical bands,
as the slingshot effect will be 3 dimensional.

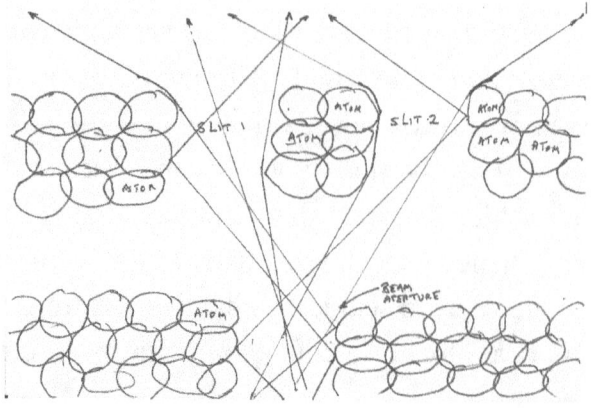

Light waves

Consequent on this explanation of the twin slit experiment, there is no need for waves to be included in the description of light.

Similarly, radio waves, gamma rays, microwaves etc can all be explained by densely packed beams of p-motes, with different separations within the beam providing the information that can be interpreted by different equipment. No wave form is necessary, unless it is generated on a screen by the receiving equipment from the information carried by the size of the gaps between p-motes.

Radio beams can be interpreted as waves because the broadcast information is by generating multiple beams of the same frequency at slightly different times, so that the receiving equipment can 'see' fluctuations in the incoming mass of beams.

How densely packed light beams can be is demonstrated by the fact that we can see stars faintly that are millions of light years from us. Although the beams are generated from an initially large area, they do tend to diverge slightly, and so having travelled millions of light years, they are still densely packed enough for us, wherever we stand on the earth's surface, to look up and see a pinpoint of light. This is our brain's interpretation of the few beams entering the area of our retina that are emanating from that star.

If the light were waves, or had a wave element, we should not see the star at its full intensity, not as a pin-point, because the waves would have decayed before then or interfered with other waves. And we know that waves decay quite rapidly and the waves from different stars would be interfering with each other, making it impossible to see a single point of light.

One has only to think of moonlight on the sea to realise the effect. We do not see a reflection of the moon, but a spread of light, broken up by the waves. Stars would look the same to us as their light travelled through the waves of light from all the other stars in the heavens.

Similarly, as described earlier, in bright light we could not make out detail as the waves would interfere with each other, as currently claimed for the pattern seen in the twin slit experiment. In fact we see more detail in bright light as more beams are reflected from the object under scrutiny.

Frequency

It is the length of separation of the p-motes in the individual beam that result in the frequency of the EMR. This is shown as a wave on a screen, but is not a wave in fact. The fact that machines measuring EMR show a graphic of a wave structure can be ignored as a proof of that structure, as these graphics are based on the old idea of lightwaves and the graphic could as easily be of a beam with gaps in, with changing intensity shown by thickness of the line rather than height of the wave and frequency by the length of the gaps rather than the wave length.

There is only p-mote/gap/p-mote. It is the interpretation of the gap, known as wavelength, that mediates our interpretation of the colour or type of EMR. and the number of beams that provide the intensity.

The 2 dimensional quality of light.

As described, light comprises 2 or more motes joined in parallel. Motes are 1 dimensional (time) so p-motes are 2 or 3 dimensional (time and width, and possibly height). The following thought experiment explains why this must be so.

A thought experiment

Imagine you are standing on the side of a street running east west, along with 1,000 other people. It is sunrise.

As the sun rises, it floods the street with light beams, all travelling east-west, across your line of vision. But you can still see every detail of the other side of the street, and so can all the other 1,000 people. How is this possible with all those light rays flooding across your line of vision?

Anyone standing at the west end of the street, looking east, would be blinded by the amount of light entering their pinpoint pupils. And how do all 1,000 people see the same detail on the same feature on the other side of the street?

The answer is that light is only 1 or 2 (spatial) dimensional. The 3rd spatial dimension (length) is (p)infinite. A p-mote has (p)infinite (equivalent to 0)) length, and travels in beams with gaps in. So there is physically nothing between you and the other side of the street (except air, which is transparent to light beams). Light beams can cross at will and never interfere with each other, or your line of sight. If they were made of 4 dimensional particles, or had a wave element, this would not be the case.

And light is composed of entities only one degree of magnitude above zero. So all photon beams contain multiple p-mote beams all diverging slightly from one another. There are also millions of photon beams from every point of the scene and they are all diverging slightly, so all the 1,000 people can see all the detail, albeit from slightly different perspectives. If the street were wide enough (Perhaps think valley, rather than street) some of the detail would be lost due to divergence of the beams, but our field of vision would be wider as beams arrive from shallower angles.

How else could we still see a star from anywhere on earth from millions of light years away? If you move a centimetre to one side you cannot still be seeing the same beam of light but you can still see the star; there must be a huge number of beams coming from it. Most have diverged enough (decayed) to no longer form photons. The pinpoint of light we see are those photons that still remain. And it cannot be a wave we see, or else it would have decayed before reaching us and interfered with all the other waves along the way.

If we look at the star with a telescope, the lenses gather in and focus more diverging beams, creating more photons, so we can see more light.

A similar thought experiment.

Imagine you are seated in a darkened cinema auditor-ium. Instead of a film a single fine but powerful laser beam is projected onto the screen. It may be possible to see a trace of the laser beam as individual photon beams are reflected from air molecules. But every one in the auditorium will see the spot on the screen where the laser beam is reflected. This means photon beams are being reflected into their eyes. In fact such a beam is passing through every part of the auditorium at very close spacings. Move your head a fraction and you will still see the spot on the screen. So that one laser beam must contain an uncountable number of photon beams. And yet, take your eyes off the spot and the auditorium looks dark. All those photons do not even light it up. They are invisible unless travelling in a direct line between screen and eye.

Now increase the number of lasers to 10, 30, 70, 100... The audience will see each spot on the screen, the number of reflected photon beams will increase a hundredfold, but the auditorium will still be dark, despite being filled with light in the form of photons travelling in every direction. Indeed, project the film and the auditorium will remain relatively dark, but each spot on the screen can be seen from any point in the auditorium.

This demonstrates that photons, made as they are of (p)motes, have no substance, due to their being only 2 dimensional.

Transparency and Reflection

P-motes travel at the speed of light, but as they enter materials their paths are altered by gravitational effects from the surrounding atoms and molecules and they are forced to twist and turn within the material. Dependant on the density, size and shape of the molecules and atoms, the apparent forward speed and direction of the beams will be changed multiple times, but the p-motes will still be travelling at the speed of light.

In some materials, the EMR is able to find a path through easily and quickly and the material is said to be transparent to, or a conductor of, that wavelength of EMR.

Sometimes the photons are broken up sufficiently to become indistinguishable as EMR or light and will emerge as undetectable individual p-mote beams or mote beams as heat radiation. They may even be burst back to motes within the material and the material will warm up. In others the path is diverted to such an extent that the beam is forced back through the same surface it entered. In the case of light we say it has been reflected. In the case of electricity and other EMR, we say the material is an insulant or non-conductive.

The frequency of the beam can also be affected, as contact with energetic atoms affects the individual p-motes in the beam. Thus a beam of white light hitting a leaf will hit molecules and atoms with particles spinning around them at certain quanta. These will bat the beam back with the frequency changed to appear green when seen.

The current theory that some frequencies are absorbed while others are reflected makes no sense as there is no explanation of what happens to the absorbed beams of light. Rocks do not glow if cracked open after millions of years in the sun, yet all the beams except those making grey are supposed to have been absorbed.

Under this explanation, most p-mote beams are reflected but at only the differing frequencies that combine to make, in this instance, grey photons. Some p-mote beams will enter the rock, but there a bursting event will mean the rock heats up from the individual motes which will eventually reach the surface causing it to feel warm to the touch, or radiate away.

A simple proof of action at a distance

Proposal:
Switch a light on in a room painted white. The white walls will reflect most of the light as it hits, and that which is not reflected cannot penetrate far, as the wall construction is an insulant to light.

After a protracted period, switch the light off. During this period light was being generated constantly in multiple beams from bursting events created in the light bulb from the action of electricity on tungsten. These individual beams will now be many millions of miles long, although all contained in the room.

Prediction:
On switching the light off, the light in the room will disappear immediately.

Observation:
I have carried out this experiment many times which confirms the forecast results. There is no time lag between the power being cut off and the room going dark.

Conclusion:

Current light theory suggests that light hitting a surface will be either reflected or absorbed. In this case, the reflected beam of photons should continue to be reflected around the room. At worst, the furnishings will absorb some light, and a few specific wavelengths may be absorbed by the wall and ceiling surface, but, energy being indestructable, and light being energetic, the beam should continue to reflect around the room and it should remain relatively well lit. If the reflector were good enough, it should continue indefinitely. But it does not. However long or short the light is switched on, so generating longer or shorter beams of photons, the instant it is switched off the beams disappear and the room goes dark.

This is equally valid with all EMR. Turn off an electric switch and the current stops flowing immediately the full length of the circuit, as for the light above.

The explanation, under this hypothesis, is that the motes making up the p-motes in the beam are spatially (p)infinite, in this case equivalent to zero in size. All they have is time and mass. Effectively, because (p)infinity cannot be measured and has no boundaries, they are spacially all one (p)infinite entity, which has only taken on units of time and mass in a finite world. As the EMR is stopped by breaking the circuit, the energy in that beam is reduced to zero so, as one, the motes must stop moving, time stops for them and they

91

lose their mass. These are the only properties they
have in a finite world. They undergo a phase transition to (p)infinity and the light disappears.

A similar example is under an infra-red lamp. As soon
as the current is switched off, stopping the bursting
events, the heat can be felt to cease. This vividly
demonstrates the relationship between motes and p-
motes; heat and light. They are formed from the same
fundamental entity.

We have seen that as light rays are generated from a
bursting event they are never exactly parallel, but tend
to diverge one from another. Their mass is too small
to allow gravity to act to keep them parallel unless
they are actually adjacent. Thus we already know that
we can see less and less detail as distance increases,
which must be because we can see fewer and fewer
photons. As we have said, photons (discernible light)
are made of many beams of p-motes. As the p-motes
travel they diverge sufficiently to become indiscern-
ible i.e. too far apart to make photons. Due to their
steady speed, we can only see photons created by a
source at a set distance from us at a set instant in time.
Reflected rays cannot combine with photons from another time instant, so the light cannot get brighter due
to more photons in a given space, and will steadily decay at the same rate as it is generated, thereby main-
taining a steady brightness.

A torch beam will only travel so far into darkness, when seen by the bearer, s the individual beams are diverging from the original bursting event, to the extent that at a certain distance they are insufficient to form photons. But when seen from the opposite direction, the torch can be seen from much further away, as the observer is seeing concentrated beams directly.

Quantum entanglement

Quantum entanglement is a physical phenomenon that occurs when pairs or groups of particles are generated or interact in ways such that the quantum state of each particle cannot be described independently of the others, even when the particles are separated by a large distance—instead, a quantum state must be described for the system as a whole.(Wikipedia Dec 2017)

An explanation of this phenomena under this hypothesis has already been provided. Because motes and p-motes are spatially (p)infinite (p-motes in length only) they are effectively one entity in the spatial dimensions, so can react as one in those dimensions. Matter particles are made of 4 dimensional c-motes, but c-motes are combined motes, the motes they are made of will do the same in all 3 spatial dimensions.

Insulation and conduction

Some materials are insulants to different wavelengths of EMR. For example, rubber and plastics are insulants for electrical currents. Lead is an insulant for x-rays.

This can be explained by the fact that all these solids are made of atoms containing c-motes at the basic level. It is the arrangement of these c-motes that decides whether a material is an insulant to a certain wavelength of EMR.

As the p-mote beam enters the surface of the material, and passes the first atom it is diverted by the atom's gravity. The extent it is diverted is dependant on the size of the nucleus and size, position and number of c-motes (electrons) which surround it and therefore the strength of local gravity. A roiling cloud of free motes (heat) will also have an ever changing effect.

It may then encounter another atom and again be diverted. The arrangement and size of the atoms, c-motes and motes that the material is made of will decide the extent and direction of the diversion.

In certain instances the combined slingshot effect will be such that the p-motes will always be ejected back through the surface, or the beam be broken up, or its frequency changed, and the material will be said to be an insulant to that frequency of EMR. In other cases, despite repeated diversions, the p-mote beam will find its way through.

Most solids are 'insulants' to light, except glass and some plastics. We treat light as a special case and say the material is opaque, but the same explanation will apply.

This also explains why light passing through a prism is diverted apparently at each face. In fact it is also being diverted slightly, at each atom, all the way through, resulting in a notionally straight path but the arrangement of the atoms is different at the surface, so the slingshot changes shape.

The splitting of white light into the spectrum can be explained by the fact that each colour in the spectrum has a different frequency pattern in the light beam, which comprises many single p-mote beams in parallel and this will cause the diversion path to be slightly different for each wavelength or frequency.

If white light is passed through a red filter, red light shows on the other side.

This is to be expected, as the plastic or glass of the filter is transparent to light, and the p-motes pick up the frequency of red from the c-motes of the dye in the filter.

If a green filter is then placed in the beam no light will pass through. This is because green dye atoms will create an insulant to red frequencies as described above. The red frequency p-motes are unable to pass through and the other side appears still green as no photons get through.

EMR conduction

Similarly some materials will enable certain frequencies of EMR to pass easily. So copper is a better conductor of electricity than some other materials as it will allow that frequency of EMR to pass through on a more direct route than lesser conductors. But it is an insulant (it is opaque) to light for the reasons explained above. Lead is an insulant to x-rays for the opposite reason.

Superfluidity

Atoms, and therefore molecules, are a seething mass of energetic motes. Motes are energy. Measuring the number of motes we interpret as the temperature of the material. Motes also act as the bonds between atoms and molecules as their light speed paths are diverted by the gravity of nearby molecules,

This means the internal gravity of the material is constantly shifting. In the same way that temperature tends to equalise, as areas become less dense free motes will tend to move into them and again be affected by denser gravities and change direction. This constant movement balances out and the material stays together as a cohesive whole.

As it is cooled, more and more of these free motes (Heat) are removed. As absolute zero is approached, so many have been removed that the links between molecules created by the shifting gravity are loosened as there are less free motes to balance it out. In the case of some liquids, this will result in superfluidity rather than solidity. Here the molecules are no longer tightly bound together by free motes, and can act individually rather than as a cohesive whole. However, the atoms are still bound as molecules, so the material is recognisable.

Eventually, as more energy is removed, the material will enter a phase transition to solid. This phase transition will happen with most materials, although some will have insufficient material left in them to complete this stage. So much energy must be removed from many materials that present equipment cannot lower the temperature sufficiently.

Superconductivity

A similar process is responsible for superconductivity. Assuming that in normal circumstances the material is a conductor, as it is cooled motes are removed and the passage between the denser atoms clears.

EMR is a beam of p-motes, or more properly a bundle of p-mote beams, as a single beam would be too small to detect. As it passes through the seething mass of motes around the atoms of the material it is passing through, it is constantly being diverted by the constantly changing gravity in its surrounding, slowing the beam's overall forward speed, although the p-motes in the beam are actually still travelling at the speed of light. Beams will also be lost from the bundle reducing the overall energy of the beam. As a consequence the carrier material has resistance.

At a certain point, as it is cooled, the density of motes between the molecules will be low enough that the beam can pass more or less directly through, without much slingshotting due to gravity, or losing beams by bursting events, and superconductivity occurs.

Magnetism

If this hypothesis is correct, and quantum mechanics and atomic particles are only mathematical constructs, then so are the strong and weak nuclear forces, which are supposed to act on those particles. That leaves magnetism, which can be demonstrated physically, not just mathematically.

We will see in Proposition 6 how the atom can be de-scribed as smaller c-motes (electrons) orbiting around a massive nucleus. At a much larger scale, this de-scribes a solenoid, where electric current, made of p-motes, spins around a much more massive core as it flows through the wire. The electric charge is positive or negative, dependant on which way the current flows. Similarly the electric current of the atom is positive or negative, dependant on which way the c-motes orbit.

There are 2 types of magnet: permanent and electro magnetic. Both generate a magnetic field. Put iron fil-ings in the field and they can be observed to form discreet orbits, emanating from the poles of the magnet, at regular spacings. At atomic level these spacings would be called quanta.

In a similar way, other magnetisable materials in the field will be attracted to or repelled from the strongest area of the field, at the poles.

Each filing is being induced as a magnet in its own right and behaves similarly to the main magnet, taking advantage of the pre-existing p-mote beams. If the magnetised filings were examined closely they would, in turn, each have their own minute magnetic field, consistent with the mass of the filing. They will clump at the poles, as they are each attracted to or repelled from the next filing in line and will tend to follow the lines of the p-mote beams as they are attracted or re-pelled by their own poles.. They lose their magnetism when they are moved away from those beams, either physically or by their own weight, in the same way as an electro-magnet loses its magnetism when the cur-rent is turned off.

Permanent magnetism

It seems strange that if magnetism is induced in a bar of iron with another magnet, it will stay magnetised, but if it is induced by an electric coil, magnetism ceases with the turning off of the current.

The assumptions of this hypothesis, as laid out previously, explain it as follows.

The field of a magnet consists of p-mote beams with a frequency that is able to pass through between iron atoms in the same way electricity, at a different frequency, will pass through also. When a permanent magnet is rubbed along an iron bar, some of its preexisting p-mote beams pass through the other iron bar and induce a flow of motes in the same direction due to gravitational attraction. These motes will be in close proximity and many will combine as p-motes. When the permanent magnet is removed, these new p-mote beams will continue to flow at the same frequency.

The apparent field is not waves or anything similar. It is these same p-mote beams affected by the gravity of the iron bar. The bigger the bar, the stronger the field. The shape of the field is dictated by the shape of the bar. As the beams emerge from the pole of the bar, they continue to be attracted by its gravity, so curve around, and are drawn to the other end in a curve, as the relative mass of the nearby bar changes along their route. As they complete the circuit, creating a continuous beam, \they continue to circle at the speed of light. The field of a horse shoe shape can be explained similarly.

Electro-magnetism

In an electro-magnet, the same description applies, but when the electric current is switched off, the p-motes in the field beams return to (p)infinity, as described for light beams when a light is turned off.

Electro-magnetism is not the same as, and has nothing to do with, electromagnetic radiation

Magnetism is not a separate force. As seen above, the only force applicable is gravity acting on p-motes.

Phase transitions to plasma & (p)infinity

If motes create gravity, the bigger the group of c-motes, the higher the gravity. I shall show in proposition 6 of this series that atoms are basically made of huge numbers of motes. We should therefore expect free motes to be attracted to atoms, and because of the gravity spin around them at discrete altitudes according to how many motes they contain. The added mass of each mote will change the possible orbit in the same way that space vehicles of a set mass and speed will only orbit a planet at a fixed height. Lower or higher and they will spiral into the planet or slingshot away from it. C-motes will try to travel in straight lines but in fact, they will be a roiling mass, as described at the big bang, due to their combined, ever

changing gravity affecting each other's paths. This
would explain the uncertainty principle. It is not just
the detectable elementary particles but also a mass
from a roiling cloud of free motes that affects the
paths of the particles around the nucleus.

This same process will lead to the well known phase
transitions from a solid to a liquid to a gas and then a
plasma as more and more motes are attracted, or are
physically added by heating, etc.. Although each mote
is spatially (p)infinite, we have seen that they cannot
occupy the same space, so more than one must have
volume. However, they can join as p-motes, and
therefore form photons, so we should expect a plasma
to be very bright as well as very hot.

Eventually, as more energy is added, the plasma contains so many motes that they can no longer move at
the speed of light. More still and they have to stop. At
that point no measurement can be made of time, as no
events are happening and there is a phase transition to
(p)infinity.

At the other end of the scale, removing motes will result in a material seeming colder. As more are removed, phase transitions occur from gas to liquid to
solid. The temperature this happens at varies with material. If enough are removed, the spaces between
fluid molecules allow the moving molecules to flow
between each other as superfluidity or for electric currents (made of p-motes) to pass directly through as su-

perconductivity, dependant on the material. Normally EMR beams have to twist and turn their way through due to slingshots by gravity around each atom, mote and c-mote. With a lack of free motes, p-motes can travel more or less straight through.

The amount of free energy in the form of motes, travelling within the material but not combined as c-motes, needing to be removed to affect a phase change will differ with each material. When all the free motes have been removed and cooling continues, the motes in the constituent molecules and atoms will start to be removed, eventually breaking apart the atoms, leaving elementary particles and the last few motes in a plasma like state. If these too are removed, at absolute zero, then nothing is left except a (p)infinity of energy.

These phase transitions require different amounts of energy for different materials, but they apply to every simple substance.

From an atomic point of view, materials become less dense with each phase transition from solid to liquid to gas to plasma, but in energy terms (motes) they become denser, and vice versa.

PROPOSITION 6
THAT THE STRUCTURE OF THE UNIVERSE IS DETERMINED BY THE PROPERTIES OF THE MOTE

Throughout the universe we see repeating patterns created by the action of locally distributed gravity from all the matter in the area under consideration.

- At the largest scale we see clusters of galaxies orbiting around a supercluster.

- The galaxies orbit around clusters of galaxies.

- The stars orbit around a galaxy's black hole.

- Planets orbit around stars,

- Moons and asteroids orbit around planets

- Electrons orbit around atoms

In between all of these, at different scales, is space debris, gases, living creatures etc., and under this hypothesis, (see propositions 1 to 3 in this series) motes, p-motes and c-motes, all of which have mass and therefore gravity, further affecting the orbits of the next smaller unit in the system.

The overall pattern tends to be flat discs arranged fairly haphazardly but with certain structures repeating, For example galaxies often have spiral arms; planets and moons often orbit elliptically. This can all be determined and forecast by known gravitational effects.

Why should these patterns change at the quantum level, for elementary particles out of which atoms and all of the above are made, as the Standard Model would have us believe?

Why all those orbits?

Gravity attracts other material things. The larger the mass of an object the higher its gravitational force, as, under this hypothesis, it contains more motes, which are the basic self-attractive building block of the universe

So, in terms of mass, a large body will attract smaller bodies to it. If they are already heading directly towards it, they will smash into it and be unable to get away again. However, if a body is approaching at an angle, 1 of 3 things will happen:

- If its speed is too great, or its mass too large, or both, it will overcome the larger object's gravity and slingshot around it and travel away again.

- If these factors are too small, its trajectory will spiral it into the larger object.

- If they are balanced, it will continue to do both, its fall towards the body balanced by its tendency to move away and it will enter into orbit around the larger object.

For any given mass and speed, there is only one condition that will enable an orbit at a particular distance around any particular mass. This is equivalent to quantum spacings at small scales. If a particle is jumping between quantum states, it, or the attracting object must be changing either mass or speed.

It follows that the orbits of the Solar System's planets and their satellites are not arbitrary, but are the only orbits each planet or moon could take given its mass and speed.

Spirals occur in larger conglomerations, as the masses of the objects already in orbit combine with the mass of the central object to encourage further approaching masses to align and take up one of the three courses outlined above. At 2 completely different scales, this explains spiral galaxies and Saturn's rings. It seems logical that the same effect will occur at all levels of physical scale. It therefore seems reasonable to infer that the orbiting c-motes in atoms will tend to do the same, just as motes' orbits create 4 dimensional c-motes..

In galaxies the central mass is a black hole.(see later) The time scale for the rotation of a galaxy is too large for modern science to be able to determine yet which of the 3 courses outlined above the objects in the spiral arms will eventually take. However, it seems logical that at some point in its spiral towards the dense centre, each object making up the spiral will reach a point at which its mass and speed will be balanced with the gravitational attraction of the galaxy as a whole, and it will maintain that particular orbit. (See Proposition 7 regarding the changing gravity of the Universe as a whole, which will also have a long term effect.)

The structure of the atom using this hypothesis.

We know from observation that the atom consists of a massive nucleus surrounded by electrons spinning in orbit. This has been observed directly, so is not purely a mathematical supposition. Let us suppose that electrons are actually c-mote clusters of fairly small size. They are attracted to the nucleus by gravity. Because of the distances involved and the weakness of gravity the nucleus must be massive compared to the c-motes orbiting it, but there is also additional mass within the orbits at any temperature above absolute zero. Motes are energy which is measured as temperature. The higher the temperature, the more free motes in any given volume. These must be taken into account when calculating the orbits of electrons (see also dark energy and dark matter). Because the quantity and movement of these free motes cannot be predicted, as they are unmeasurable being spatially (p)infinite, the orbits are unpredictable (Heisenburg's uncertainty principle) and the electrons will change orbit (quanta) with changes in local gravity. Local gravity will be constantly changing unpredictably as all the free motes and the c-motes in the nucleus constantly sling-shot around each other.

It is known that the higher the temperature, the more energetic atoms are. This is normally envisaged as the atoms vibrating in some way. It is actually, if this hypothesis is correct, all those extra motes roiling around at the speed of light due to minute local gravitational effects.

So in any atom we have a fixed mass of c-motes in the nucleus, combinations of c-motes, equivalent to the elementary particles (electrons) orbiting the nucleus, and a random number, changing with temperature, of spatially (p)infinite motes roiling around the free spaces in between.

Compare this to what we know of the Solar System. We have a large central mass, the sun; orbiting smaller masses, the planets; and in between, asteroids, comets, solar winds etc, roiling around randomly.

(Actually, not quite randomly. They are affected by constantly changing gravitational effects, making their paths difficult, if not impossible, to calculate) This pattern repeats at whatever cosmic scale is considered.

Heisenburg's uncertainty principle

The motes within the orbiting material and within the nucleus are travelling at the speed of light in haphazard directions, being pushed and pulled by locally speedily changing gravities. Because these fluctuations are so rapid, the orbits of the surrounding c-motes (electrons) will be constantly changing. Therefore they cannot be predicted with any certainty. C-motes will change orbit by a fixed amount (quanta) when additional motes join the c-mote cluster. As motes are all identical, their mass must change the orbit by a fixed amount. It is not due to the matter/wave nature of all quantum objects, as, under this hypothesis, this does not exist and is not necessary for a GUT.

Observation must, of necessity, change the state of any particle observed. We have seen that the basic unit of the universe is the mote, a 1 dimensional entity of (p)infinite size. A p-mote is 2 or 3 dimensional, but cannot be observed on its own. A 4 dimensional photon, which is necessary to make an observation, is many times this size. To observe with a photon of whatever frequency of EMR, it has to move from its point of generation to the quantum object being observed and be reflected from the object to our equipment. The mass of a photon smashing into a subatomic particle, let alone a basic c-mote, has to change the composition and/or position and direction of the object under observation. Normally this does not matter, but at quantum scales it does. In scale it could be likened to the sun smashing into the Earth so that some being in another galaxy could observe the Earth. Not a very useful thing to do.

We have already seen (Proposition 3 of this series) that a photon cannot be massless, and is 4 dimensional so mathematical arguments against this logical proposition cannot apply.

Pauli's exclusion principle.

'The answer to the question of what stops matter collapsing in on itself, surprisingly, was not proven until 1967, when physicists Freeman Dyson and Andrew Lenard showed that the stability of matter is down to quantum mechanical effect called the Pauli exclusion principle......... this means you can't pile lots of (fermions and bosons) into the same place. This is the reason why atoms are stable and chemistry happens'. (Brian Cox; *Wonders of the Universe by Brian Cox and Andrew Cohen*, pub: BBC & William Collins)

This indicates that the particles would all crush into the lowest possible orbit without it. There could then be no chemical reactions and matter as we know it could not exist. (Précis of Brian Cox; *ditto)*

Do the planets and moons, or Saturn's rings, even stars, crush into the lowest possible orbit around their star/planets/galaxy? This principle must also apply at all cosmological levels, or it is formulated on the wrong data.

If fermions and bosons were to exist physically, rather than mathematically, their orbit would be determined by speed, mass and gravity; nothing else. I have shown above that the orbital path is solely due to the mass and speed of the 2 objects. This applies at sub-atomic as well as cosmic scales.

Quanta

These above orbits are also manifestations of quanta. That is, the only possible orbit for an object of a given size and speed around a given mass. Change any of these by a fixed amount and the object must change orbit by a fixed amount (quanta) or else face absorption or ejection.

The Nucleus

We have already seen that a mote takes up no space, but has mass, so is attractive (has gravity) when another mote is adjacent; and a c-mote is a combination of 2 or more adjacent motes spinning around each other at the speed of light. All of space is filled with 1 dimensional motes moving on haphazard paths at the speed of light, being constantly diverted by the gravity of nearby motes as well as other more massive bodies. Hence the CMB. When 2 motes on a collision course combine as a c-mote a 4 dimensional spinning entity is formed. This locally has increased the gravity field so will tend to attract more motes, increasing the local gravity again. In this way, a vast mass of motes will build up in a small volume. Because of the increased gravity in this small volume, the motes, although not all combined will tend to spin around in a ball, similar to a sun but on a sub-atomic scale, made of motes and c-motes rather than hydrogen and helium atoms. This is possible as the mote is spatially infinite, so billions could fit in this space. This roiling mass will not all be spinning in the same direction, and some additional c-motes will form, although the overall ball will tend to spin in one direction, just as other structures do at cosmic scales, although made of different materials whose atoms spin in all directions.

Localised concentrations will occur, just as atoms cluster in matter, matter clusters in star systems, stars cluster in galaxies, and galaxies themselves cluster. It may be these concentrations in the atom that science is calling elementary particles.

At a certain size, no more motes will be able to join, just as stars and galaxies are of different sizes, but fit in a definable range. The size will differ dependant on how many c-motes have joined the nucleus which will affect the overall mass. The nucleus is now massive enough to attract the 'planetary' c-mote electrons and keep them in orbit at quantised spacings. The atom is formed but motes still are able to move within and around it, quantity dependant on temperature, just as comets and other objects move in and around star systems. They will not join if their paths, speed and distance are sufficient to prevent it.

Analogously, comets, asteroids and similar space debris may crash into planets occasionally, but mostly will circle around inside a star system, pulled by various gravities, without affecting the system as a whole.

Cosmic patterns repeat

This description of the atom reflects very well a star system, of a large central mass with orbiting planets or a galaxy with stars orbiting around a dense black hole. Also what is known about the structure of a neutron star, with its fizzing electrons around a dense nucleus, confirming that patterns repeat whatever the cosmic scale.

It does not need exotic particles or special forces to be created for it to work. It is all down to the self-attraction (gravity) of motes.

Other repeating patterns can also be discerned. Weather systems and water down a plughole can be likened to the swirl of galaxies. The stars correspond to the molecules of air or water, and no edge can be measured, as more are drawn in all the time. The twister at the centre of an extreme weather system can be likened to the black hole of a galaxy as the spiralling entities reach the centre and are sucked in. The twister breaks up as it hits the ground into chaos. The black hole has nothing solid to hit, so compresses to (p)infinity.

The galaxies are moving chaotically as gravity is constantly changing due to this movement, causing more swirling. The time scales are too large in the Universe, but again can be likened to a weather system which is constantly evolving. Because of this constant changing, and because we cannot see the edge of the universe, we cannot calculate the centre of the universe, where the Big Bang happened, but there logically must be one, from which all the Universe originally expanded.

The need for additional forces to that already known to act throughout the universe, gravity, is entirely eliminated. The 3 extra forces claimed for the atom are not known in larger universal constructs and are only there to make the mathematical formulae work. As for the 17 different particles, they may or may not exist. It is possible that they are large c-motes under different names.

The same structures occur at all scales, all due to gravity alone, throughout the universe. We see galax-

ies around black holes, planets around stars, moons
and rings around planets and electrons around the
nucleus. Galaxies spin, star systems spin, atoms spin
and c-motes spin. It seems likely that this would all be
due to the momentum of all those motes spinning
around c-motes, the very basis of matter.

Natural selection

This description of the formation of the atom will result in entirely arbitrary chemical elements being
formed as individual atoms. That we normally find
them conglomerated together is due to the elements
coming from space debris from exploding stars, where
similar atoms are being manufactured under similar
conditions within a confined volume, so that certain
types are likely to be successful. This is a similar
concept to biology's natural selection.

Caucasians, indo-chinese and africans are all recog-
nisably human. Only their DNA differs due to natural
selection.

Similarly, there is no reason to believe that iron found
on earth will have an identical atomic structure to that
found in, say, the Orion Nebula. It may be, for instance, a completely different colour, but still identifi-
able as iron.

Nuclear fusion and fission

It also explains why so much energy is released in nuclear explosions. The materials are made entirely of energy (motes) locked in its atoms. Fusion and fission are no more than bursting events massive enough to destroy the nucleus, releasing billions of motes from each atom. Where are the quarks and fermions that should be the result of these explosions if the standard model is true? All we see is heat and EMR (motes and p-mote beams), which is exactly what should occur if this hypothesis is true.

The Cosmic Microwave Background

This is usually interpreted as the left over radiation from the Big Bang. In a way it is, as it has always (in terms of this universe) existed, as have all the motes in the universe. However, I believe it is actually a photograph of dark energy, being all those motes roiling around the universe, undetectable except as heat when hitting us or our apparatus. A photograph or other record of EMR, including the CMB, can only detect those motes hitting the apparatus. The free motes in dark energy are not in beams, so the variations in apparent wavelength in the photo are the result of the motes roiling around each other due to their mass and constantly changing local gravity.

Ratios of dark matter, light and matter

The above hypothesis has several consequences:

1) It explains the ratio of light to matter as the likelihood is that motes will join much more easily as p-motes than c-motes, when they are close to and escaping bursting events, rather than crossing each others paths randomly, so we should expect the universe to be filled with light, with much less matter. However, the most likely thing is that motes will not get close enough to others to join at all, so will just stay as motes, detectable as heat.

2) Given that the most abundant thing in the universe is motes, which have mass, and are energy, but have no finite spatial dimensions, this explains the presence of so much dark matter, and dark energy. They are manifestations of the same thing, a universe filled with spatially infinite but energetic motes, which have mass and therefore gravity.

3) If the Universe is filled with motes, we should expect to be able to detect them somehow. We can. We commonly detect energy as heat. The more motes present, the more heat we feel. However, we only detect those motes directly affecting us, i.e. actually touching our bodily receptors or equipment, so we should not expect the search for dark energy or dark matter to find them in outer space by detecting particles of some kind. (see above re CMB) Although motes are what give rise to all 3 manifestations of energy and mass (heat, light and matter) they have to touch us or our equipment to be detected.

Dark matter/energy in outer space can only be detected by its affect on other, material cosmological subjects, created by its mass and movement. If it touches us, we recognise the motes it is composed of as heat, but when not touching us or our equipment, it is dark matter/energy, the mass of which has gravitational effects.

As an analogy, our eye can only see those p-mote beams actually reaching our retina. All other light is unseeable to us but we know it exists throughout the universe. We could say it is 'dark light'.

Black Holes

We know that a black hole is an area of immense gravity. This must mean that it has an immense mass and tends to draw more matter and light in all the time. Because the force of gravity is so great due to that overwhelming mass, even light is unable to escape.

This must also mean that its gravity is overcoming the energy of motes. If Motes cannot escape either they also will be drawn in, so the area around must be getting colder as well as duller and less dense.

Because we are used to seeing diagrams of black holes showing the light cone as being long, narrow tubes, this is how we (I, at any rate) tend to think of them. This is not logically accurate, as they must draw matter from all directions into the centre, as gravity acts in all directions, so they must be spherical. But what happens at the centre as more and more matter is drawn there?

More and more material is dragged towards the centre due to the strength of gravity emanating from there. As the density gets higher, and the mass of matter pressing towards the centre gets larger, the pressure would build as more and more motes, p-motes and c-motes were packed more densely together in their constituent matter with the orbiting electrons being the first to be forced into the nucleus, destroying the structure of the atom and matter as we know it, releasing motes. Normally this would be a bursting event, but the pressure would prevent any bursting.

Current theory would say that as the heat energy is released, the temperature would exponentially rise. However, this is another oxymoron, as there is no such thing as heat energy. Heat is a measurement. It has to have units to be meaningful. The measurement of temperature is only our interpretation of the energy of the number of motes contacting ourselves or our equipment. As we would not be there, and it would not be radiating, this observation is meaningless.

At a certain point, when all the matter and light is compressed into a very small volume, c-motes would be at a standstill, their motes still spinning around each other, and p-motes and free Motes would not be able to travel very far in a straight line at the speed of light. Gravity would pull them into tighter and tighter curved paths.

This is the same as nuclear fusion, but the force of gravity would be so strong that no bursting event could take place and the energy could not radiate as in a star. Essentially a plasma contained by its own gravity.

There would be an almost solid volume of motes, more dense than a neutron star.

Eventually, the motes could not combine or travel, due to gravity pulling everything more and more closely together; volume would be close to zero and the motes, previously being independent units of energy, would start to merge, this being their final possibility for movement. A plasma would briefly appear, which is the densest state of energy we know in this universe. On ceasing to be able to move, the final mass of motes would change state into (p)infinity as a phase transition occurs.

Matter per se would disappear, along with spatial dimensions. Motes would all be packed into a single (p)infinite point. As it is dependant on the movement of energy, mass would disappear, destroying gravity. If nothing could move, having no dimensions to move in, time would stop as it could no longer be measured. A singularity occurs. In other words, (p)infinity would re-occur /continue. This does not mean the matter in the black hole moves to somewhere outside this universe. Outside/inside have no meaning where there are no boundaries. Infinity just is.

This process would be continuous as matter was drawn towards the massive centre by gravity. Once there it effectively just disappears as motes change phase state to (p)infinity

Here the deconstructed matter will stay as (p)infinite fundamental energy until the hyperbolic cycle comes round again and another phase transition will create time and another universe (??) That last sentence has lots of oxymorons about time, which does not exist in (p)infinity, but we have to be able to imagine it somehow!

The Expansion of the universe.

We tend to think of expansion as being like a balloon being blown up. However, the universe has no skin. As we have already seen, when talking of infinity in proposition 1, the edge of the universe is free to expand into a void. This means that, unlike a balloon where the density of air increases as it is inflated, the density of motes decreases as the Universe expands. As it expands, more motes are needed at the surface to maintain the expansion. Internal density decreases, and therefore overall gravity also decreases ,At the surface the motes, previously pulled around by ever changing gravity, slowing their 'as the crow flies' progress, can travel in straighter and straighter lines directly outwards The resulting increase in volume is hyperbolic.

The common diagrammatic model of Lambda-CDM expansion shows a 2 dimensional view of the growth of the universe. It starts from a singularity, but due to the limitations of diagrammatic modelling, only shows expansion in one direction. As is normal in explosions, expansion would be expected to continue in all directions. It is the speed of expansion over time that is interesting in this diagram, (*copied from Wikipedia Jan 2018*):

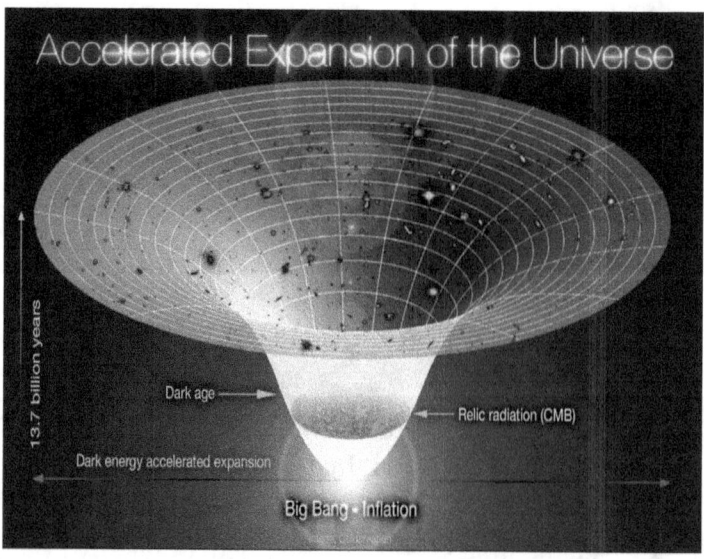

The model shows early inflation, followed by continuing but slowing growth but this appears to be speeding up again over the last few billion years. This indicates that expansion is following a hyperbolic curve.

Hyperbolic curves are time limited and move from and to infinity within those time boundaries.

This can also be logically forecast as follows:

As the galaxies move away from each other at a faster and faster rate, the gaps between the galaxies become bigger and bigger, and gravitational forces therefore get smaller, inversely proportional to the square of the distance apart. The total mass of the universe remains the same, apart from a small fraction disappearing into infinity in black holes, so the gravitational force binding the universe together must therefore get relatively smaller as the universe gets bigger as time goes on. In turn, expansion can be expected to speed up, as it is. Time will speed up relative to our current observations as the universe gets less dense, as per special relativity, so from today's perspective, this could all happen relatively slowly, but will be quicker measuring at a moving time scale.

If the galaxies are expanding, so will everything else within the galaxies, due to ever weakening gravity in the universe as a whole. Stars will get bigger and engulf their planets or explode. If all universal matter becomes less dense, even spaces between molecules will get bigger; spaces between nuclei and electrons will get bigger, until eventually gravity cannot hold them together. Everything will disintegrate into a huge cloud of c-motes and p-motes, equivalent to a phase transition to a plasma, although no additional energy will be involved. In fact the energetic content within all matter will reduce, as motes try to balance their density, flowing into less dense areas. although the overall energetic content of the expanding universe will stay the same. If my proposal of the mote as the fundamental entity is correct p-motes and c-motes will eventually break apart. By the time everything returns to motes, the hyperbolic expansion will have increased to such a speed, (m)infinite – not quite truly (p)infinite, that density and the temperature of space will reduce to near absolute zero and the motes will jump to a (p)infinite energy state in a phase transition, just as they were before the universe, and the universe will be no more.

The expansion of the universe can be likened to the phases of matter. At the Big Bang, from **(p)infinity** a hot **plasma** erupted. In the early stages it was a dense cloud of motes, as a **solid** is a dense mass of atoms.

As expansion progressed, a phase change occurred to **liquid**; less dense but c-motes and p-motes were appearing, and their ever changing field of gravity was making them roil about. Further expansion and as things got less dense, a phase transition to **gas** occurred. This is where we are now, with a few solids (equivalent to atoms in a gas; the galaxies being equivalent to molecules.) within a vast space of nothing but motes (dark matter), but virtually no 4 dimensional matter. With hyperbolic expansion we can expect the next phase to be another cold **plasma** as matter disappears with the reduction in gravity, until only energy is left following a further phase transition an unmoving **(p) infinity.**

PROPOSITION 7
THAT A GRAND UNIFIED HYPOTHESIS IS POSSIBLE BUT NOT A GRAND UNIFIED THEORY, DUE TO THE REQUIREMENT TO INCLUDE INFINITY

About this model

"A model is a good model if it:

1) Is elegant

2) Contains few arbitrary or adjustable elements

3) Agrees with and explains all existing observations

4) Makes detailed predictions about the future "

Stephen Hawking & Leonard Mlodinow *The Grand Design* (Published by Bantam Press 2010

Quantum Mechanics and The standard model.

For the last several hundred years physicists have been getting deeper and deeper into mathematical proofs rather than empirical proof. This has been ne-cessary as we have looked at smaller and smaller structures. Physicists now routinely mathematically examine structures which we cannot physically ob-serve. They seem to have gone wrong somewhere along the way. We saw in proposition 3 of this series how mathematics can model unreality and this seems to have happened with quantum mechanics.

We now have the situation where they openly state that if you understand it, you don't know about it; where particles can have no mass and still be said to exist, despite $E=mc^2$; where particles can pop out of nowhere and then disappear again; where cats can be alive and be dead at the same time (possibly)... the list of absurdities goes on, and they would have us be-lieve it all as they have 'proved ' it mathematically.

The formulae may be mathematically correct, but does that, of itself, mean they are true?

We started with the discovery by Rutherford that the atom was composed of a nucleus surrounded by smal-ler bodies. This was observed by diffraction, which of itself is suspect if the hypothesised description of light above is correct. But I can accept that it was observed directly, whatever the method, and suggest that the observed electrons could be the c-motes of this hypothesis.

However, to the mathematicians, gravity did not seem right somehow, so it was proposed that the electrons were held in orbit by photons. Only trouble was that they needed to be zero mass and therefore zero energy for the formulae to work. Photons travel at 300,000 kph. How can they have zero energy, and at that acceleration around the nucleus have zero mass? If true, why does a star lose mass over time when all it gives off are light and heat? This completely contradicts other formulae, which can be shown by direct observation to be true.

However, the atom's nucleus could not be studied by observation, so maths continued to be used.

It was assumed that the atom, being 4 dimensional, was made of 4 dimensional particles so calculations were put in hand to find them. Sure enough, some were found, protons and neutrons, and the figures were convincing until it was noticed that gravity would not contain them. So a new force was invented. Its only basis, it seems to me, was that it would hold the supposed nucleus together. Then someone asked how the force worked, and gluons were invented, and the formulae were adjusted to allow it. However, they

showed that it only worked if gluons were also zero
energy and mass. What happened to $E=mc^2$? Substituting, the speed of light $=0$! And if they have no energy, how do they act?

Then it was found something was still missing, so
they came up with the Higgs Boson. This has no
known properties but has been observed (apparently).
Only trouble is the observed particles were so small
that they could not be directly observed so the collision results had to be computer simulated to be dis-
played on a screen. And I bet they used the existing
formulae in their algorithms. A circular argument.
Look for something by rehashing the same formulae
that did not work in the first place, and you are bound
to find it.

As for quarks and their 6 flavours, muons, tau and
neutrinos, in total 17 different particles (plus various
'flavours') and 4 forces, they all seem to have been
invented to 'prove' a previous formula that was not
quite working properly. It is all circular arguments.
There is no actual evidence, or alternative mathematical description, to verify their existence, only the fact
that this mathematical description seems to fit with
'observation by calculation', to a great extent (but not
exactly). Such observations being carried out by computer also programmed using the same mathematics,

To say that it works empirically is no proof either. I
could say when I turn a tap on water will come out,

and I could measure its temperature and flow, and predict it will spill all over the kitchen floor, then carry out the experiment and prove it. But that does not prove where the water comes from or how the plumbing is arranged. Just to say the formulae work is not proof that the suppositions behind them are correct.

And the biggest stumbling blocks are:

- How did the Big Bang contain the universe, with its near (m)infinite number of 17 (or more) different 4-dimensional particles in a singularity of no dimensions?

- Where did those particles and forces come from? And why were the forces ready and waiting to act on them?

- Those 2 extra forces needed to mathematically construct the atom. They exist nowhere else in the structure of the universe.

- Zero mass particles, who's energy must also be zero.

Only the hypothetical alternative universe described above can resolve these and similar issues arising from a mathematically described universe.

TABLE OF FUNDAMENTAL ENTITIES

No. of dimen-sions	Dimension type	(p)infinite/finite	Possible entity	Observed as
0	none	(p)infinite	Energy	Unobservable
1	time	(p)infinite in 3 di-mensions	mote,	heat(when en masse)
2	Time + 1 spatial	(p)infinite in 2 di-mensions	P-mote	Light & EMR (also only when en masse)
3	Time + 2 spatial	(p)Infinite in 1 di-mension	P-mote or [rarely] c-mote	light or [rarely]matter
4	Time + 3 spatial	finite	C-mote	matter

SUMMARY

We can summarise what has been deduced in the first 7 propositions of this series as follows:

- Mathematical infinity ((m)infinity) and physical infinity ((p)infinity) are not the same.

- Physical infinity ((p)infinity) is a phase state of energy, which has no boundaries and where nothing moves. It follows that it can have no spacetime, and if nothing can be measured, maths cannot be applied. However, it can be logically and philosophically defined.

- The fundamental entity of the universe can be elegantly and logically envisaged as the mote, which is spatially (p)infinite, but has 3 attributes: mass, time and energy.

- Energy in the form of Motes in a finite state, appeared at the Big Bang by a phase transition from (p)infinity. They accelerated to the speed of light, acquiring mass and creating time and the spatial dimensions.

- Motes en masse can be detected as heat, dark matter and dark energy.

- Motes can join by gravitational attraction if on a near parallel path (a p-mote) which can be 2 or 3 dimensional. The 4th dimension, length, is (p)infinite explaining action at a distance.

- P-motes are the basis of EMR including the light spectrum.

- P-motes have mass and width but (p)infinite length. When in beams, we detect the energy and we interpret the spacing between them as EMR. Frequency depends on distance apart of the individual p-motes in the beam.

- Light needs no wave component when described in terms of p-mote beams.

- The photon is the smallest detectable bundle of p-motes.

- Motes can similarly join if on converging paths, and spin around each other in 4 dimensions as combined motes (c-motes).

- C-motes are the basis of matter.

- As the combined motes are spinning in a very small space at the speed of light, they are effectively solid.

- Motes join because they are self attractive, which we call gravity. The more motes in a given space, the higher the force of gravity.

- The mote is the fundamental unit of the universe, and gravity is the fundamental force.

- We can detect motes directly as heat, and indirectly as dark energy and dark matter. The CMB is also the detection of motes.

- The atom is composed of a nucleus of very large mass (a huge number of c-motes) with smaller c-motes spinning round it, just as planets orbit stars. This structure repeats at different scales throughout the universe.

- Quanta are discrete orbits dictated by the speed and size of the orbiting c-mote and the nucleus just as a planet's orbit around a star is related to their respective speeds and masses. The uncertainty principle is explained by the roiling mass of motes rapidly changing local gravity within the atom.

- There is no requirement for the multiple particles and forces envisioned by physics and 'proved' by the scientific method, requiring illogicalities such a zero mass particles and cosmological constants etc inserted just to make the equations balance.

- Heisenberg's uncertainty principle, Pauli's exclusion principle, quanta, black holes, electromagnetism, superfluidity and superconductivity, insulation et al, can all be explained simply in terms of the mote and gravity.

- Newtonian and quantum physics are unified, thereby presenting a very simple and elegant model for a GUH (Grand Unified Hypothesis).

By hypothesising the mote, we are able to construct a logical and elegant Grand Unified Hypothesis which overcomes the problems created by the mathematical descriptions of the standard model. Unfortunately, it is not amenable to mathematical proof as infinity is involved so must currently remain a hypothesis, not a theory.

This hypothesis allows conformity at the sub-atomic level with the larger, repeating patterns seen throughout the universe created by gravity, and eliminates the necessity for other forces to be introduced.

It also provides a neat, circular scenario for energy, matter and light creating the universe by a simple phase transition at the Big Bang and describes how it can exit via black holes and will finally disappear into another phase state as the universe ends in (p)infinity due to its expansion following a hyperbolic path. Matter may disappear, but energy in the form of motes will only phase transition to a new state, physical infinity.

However, our maths is designed to solve the problems of a 4 dimensional universe, and the mote is a 1 dimensional entity, so it seems unlikely that this hypothesis can be proved by the Scientific Method of mathematics as we know it.

This does not, of itself, make it wrong, only that a new maths will be needed where physical infinity can be accepted.

.

Jan Venters
March 2023